Charge Transfer in DNA

Edited by
Hans-Achim Wagenknecht

Further Reading from Wiley-VCH

L. V. Yakushevich

Nonlinear Physics of DNA

2nd Edition

2004

ISBN 3-527-40417-1

C. M. Niemeyer, C. A. Mirkin (Eds.)

Nanobiotechnology

Concepts, Applications and Perspectives

2004

ISBN 3-527-30658-7

B. L. Feringa (Ed.)

Molecular Switches

2000

ISBN 3-527-29965-3

V. Balzani et al. (Eds.)

Electron Transfer in Chemistry

2001

ISBN 3-527-29912-2

Charge Transfer in DNA

From Mechanism to Application

Edited by
Hans-Achim Wagenknecht

WILEY-VCH
WILEY-VCH Verlag GmbH & Co. KGaA

Editor:

PD Dr. Hans-Achim Wagenknecht
Technical University of Munich
Department of Chemistry
Lichtenbergstrasse 4
85747 Garching
Germany

Library of Congress Card No.: applied for
British Library Cataloguing-in-Publication Data:
A catalogue record for this book is available from the British Library

Bibliographic information published by Die Deutsche Bibliothek
Die Deutsche Bibliothek lists this publication in the Deutsche Nationalbibliografie; detailed bibliographic data is available in the Internet at http://dnb.ddb.de.

Composition ProsatzUnger, Weinheim
Printing betz-druck GmbH, Darmstadt
Bookbinding Buchbinderei J. Schäffer GmbH, Grünstadt

Printed in the Federal Republic of Germany
Printed on acid-free paper

ISBN-13: 978-3-527-31085-2
ISBN-10: 3-527-31085-1

Foreword

Hans-Achim Wagenknecht has put together a terrific book on distant charge transport in DNA! This area of science has grown enormously since the publication of the Barton-Turro paper in Science in 1993. I well remember reading the paper, puzzling over the remarkable ability of DNA to facilitate long-range charge transfer. My laboratory had been engaged in investigations of electron transfer in proteins since 1972, including an experimental demonstration of rapid long-range electron tunneling through cytochrome c in 1982, and naturally I was interested in comparisons of DNA with protein in mediating distant electronic couplings. The DNA work was really fascinating, albeit controversial, as there was heated discussion that centered on the wirelike conductivity of DNA as well as on other issues related to the efficiency of distant charge transfer chemistry. Thankfully, as Hans-Achim has so clearly pointed out in his introductory chapter, many of the disagreements have been resolved through really heroic efforts of both theoreticians and experimentalists over the last ten years or so. This book, which I can enthusiastically recommend, nicely presents the picture of this very exciting area of science at this critical point in time.

The roles of superexchange coupling, hole transport, and electron injection in different modified DNAs are treated extensively in several chapters. In Chapter 2, O'Neill and Barton discuss distant oxidative damage and the role of sequence as well other factors in the efficiency of distant charge transport. In Chapter 3 Carell and von Meltzer discuss electron transfer repair of damaged DNA, and in Chapter 4 Lewis and Wasielewski provide a thorough treatment of DNA charge transport energetics and dynamics. Hole hopping through bases is reviewed by Kawai and Majima (Chapter 5) and by Dohno and Saito (Chapter 7), reductive events are discussed by Rokita and Ito (Chapter 6), and the spectroscopic signatures associated with guanine oxidation are treated by Sharirovich and Geacintov (Chapter 8). The volume is wrapped up beautifully by Hans-Achim and his colleague Torsten Fiebig with a discussion of ultrafast DNA dynamics and other fundamental mechanistic issues.

Remarkably, every chapter includes citations to papers published in 2004! And the book is filled with references to papers that are no older than a couple of years.

Charge Transfer in DNA: From Mechanism to Application. Edited by H.-A. Wagenknecht

ISBN: 3-527-31085-1

All theorists and experimentalists interested in charge transport phenomena will want to have "Charge Transfer in DNA" in their personal libraries. Congratulations not only to Hans-Achim but also to the publisher (led ably by Elke Maase) for producing this scholarly work in timely fashion.

May 2005

Harry B. Gray
California Institute of Technology, Pasadena

Preface

The idea that DNA might serve as a pathway for charge migration was suggested over 40 years ago by Eley and Spivey. The interest in DNA-mediated charge migration grew in the scientific community over the subsequent decades. Significant experimental improvement came with the development of automated DNA phosphoramidite chemistry, which gave access to modified DNA systems bearing charge donors and acceptors covalently attached to the oligonucleotides. Over the last 15–20 years, the fundamental question of whether DNA represents such an unique medium for long-range charge transfer has been discussed in a highly controversial fashion. Remarkably, DNA was considered to be a molecular wire, semiconductor, or insulator. Motivated by the biological relevance, the excitement about this topic grew enormously. Working groups from a variety of chemistry subdisciplines, such as organic chemistry, inorganic chemistry, physical chemistry, and biochemistry, as well as biologists, physicists, and material scientists have contributed significantly to this research topic. Still today, this interdisciplinary character represents an important and exciting aspect of this subject.

The initially extreme controversy has been solved by the description of different mechanistic aspects and their experimental verification. It has become clear that DNA-mediated charge transfer processes can occur on an ultrafast timescale and can result in reactions over long distances. Initially, DNA-mediated charge transfer was the subject of considerable interest, with biological relevance in the formation of DNA damage that can result in severe consequences such as mutagenesis, apoptosis, or cancer. Charge transfer now plays a growing role in the development of DNA chips detecting single-base mismatches or various DNA lesions by electrochemical readout methods. Most recently, several groups have started to elucidate how DNA can be modified in order to enhance its electron transfer capacities for new nanotechnological devices based on DNA or DNA-inspired architectures.

This book contains nine articles covering the most important aspects of charge transfer processes in DNA. Additionally, the first chapter is intended to be a short overview, written in textbook fashion in order to provide a clear introduction. I am very grateful to all the authors for their contributions to this book. Finally, I would

Charge Transfer in DNA: From Mechanism to Application. Edited by H.-A. Wagenknecht

ISBN: 3-527-31085-1

List of Contributors

Jacqueline K. Barton
California Institute of Technology
Division of Chemistry and Chemical Engineering
1200 East California Boulevard
Pasadena, CA 91125-7200
USA

Thomas Carell
Ludwig-Maximilian-Universität München
Fakultät für Chemie und Pharmazie
Butenandtstr. 4
81377 München
Germany

Chikara Dohno
Kyoto University
Department of Synthetic Chemistry and Biological Chemistry
Faculty of Engineering
Kyoto, 615-8510
Japan

Torsten Fiebig
Boston College
Chemistry Department
Chestnut Hill, MA 02467
USA

Nicholas E. Geacintov
New York University
Department of Chemistry
29 Washington Place
New York, NY 10003
USA

Takeo Ito
Kyoto University
Department of Energy and Hydrocarbon Chemistry
Kyoto Daigaku Katsura, Nishikyo-ku
Kyoto, 615-8510
Japan

Kiyohiko Kawai
Osaka University
The Institute of Scientific and Industrial Research
Mihogaoka 8-1, Ibaraki
Osaka, 567-0047
Japan

Frederick D. Lewis
Northwestern University
Department of Chemistry
Evanston, IL 60208
USA

Charge Transfer in DNA: From Mechanism to Application. Edited by H.-A. Wagenknecht

ISBN: 3-527-31085-1

Tetsuro Majima
Osaka University
The Institute of Scientific
and Industrial Research
Mihogaoka 8-1, Ibaraki
Osaka, 567-0047
Japan

Melanie A. O'Neill
Simon Fraser University
Department of Chemistry
8888 University Drive
Burnaby, British Columbia, V5A 1S6
Canada

Steve E. Rokita
University of Maryland
Department of Chemistry
and Biochemistry
College Park, MD 20742
USA

Isao Saito
Nihon University
NEWCAT Institute
School of Engineering
Tamura
Koriyama, 963-8642
Japan

Vladimir Shafirovich
New York University
Department of Chemistry
29 Washington Place
New York, NY 10003
USA

Martin von Meltzer
Ludwig-Maximilian-Universität
München
Fakultät für Chemie und Pharmazie
Butenandtstr. 4
81377 München
Germany

Hans-Achim Wagenknecht
Technical University of Munich
Department of Chemistry
Lichtenbergstrasse 4
85747 Garching
Germany

Michael R. Wasielewski
Northwestern University
Department of Chemistry
Evanston, IL 60208
USA

1
Principles and Mechanisms of Photoinduced Charge Injection, Transport, and Trapping in DNA

Hans-Achim Wagenknecht

1.1
Introduction

The first observation that the π-stacked array of base pairs in B-form DNA might serve as a pathway for charge migration was published over 40 years ago [1]. Since then, the basic question of whether DNA serves as a wire or conducting biopolymer for long-range charge migration has been discussed in an intense and highly controversial scientific dispute. DNA was considered to be a molecular wire, a semiconductor, or an insulator [2]. Barton and coworkers pioneered this research through remarkable contributions about photoactivated charge transfer chemistry in DNA [3]. Motivated by the biological relevance of DNA damage and also by the controversy about charge transfer in DNA, interest in DNA-mediated charge migration grew enormously in the scientific community in the 1990s [4]. Research groups from different chemistry subdisciplines, such as organic chemistry, inorganic chemistry, physical chemistry, and biochemistry, as well as biologists, physicists, and material scientists have contributed significantly to this research topic. This interdisciplinary nature represents an important and exciting aspect of this subject.

Based on these experiments and results, a clear picture about charge transfer processes in DNA has emerged by now. The extreme controversy has been solved by the description of different mechanistic aspects, mainly the superexchange and the hopping mechanisms, which have been verified experimentally [5]. It has become clear that DNA-mediated charge transfer can occur on an ultrafast time scale and can result in reactions over long distances [4]. Hence, DNA-mediated charge transfer has been the subject of considerable interest, having biological relevance in the formation of oxidative damage to DNA that can result in severe consequences such as mutagenesis, apoptosis, or cancer [6]. Additionally, charge transfer plays a growing role in the recent development of DNA chips or microarrays detecting single-base mismatches or various DNA lesions by electrochemical readout methods [7]. Moreover, knowledge about charge transfer processes in DNA can be used for nanochemical applications, such as DNA-based devices [8].

Charge Transfer in DNA: From Mechanism to Application. Edited by H.-A. Wagenknecht

ISBN: 3-527-31085-1

1.2
Synthetic DNA-Donor-Acceptor Systems

Due to the short lifetimes of the natural DNA bases, it is necessary to modify oligonucleotides with suitable chromophores as tools in order to photoinitiate and study charge transfer reactions in the DNA helix. The initial difficulties in the synthesis of oligonucleotides that are covalently modified with suitable charge donors and acceptors, as well as the high concentrations necessary for the measurements of charge transfer rates by laser spectroscopy, were reasons for the first charge transfer experiments to be performed with noncovalently bound intercalators [9]. Using these assays, it is difficult to determine how the base sequence influences the charge transfer efficiency. More importantly, these experiments provide only very limited information, since the distance between the unspecifically bound charge donor and acceptors is unknown. The major concern with such assays is that the cooperative binding of donor and acceptor could provide the structural basis for a shortcut yielding an efficient short-range charge transfer and fast electron transfer rates.

Thus, new DNA systems that bear the charge donor and the charge acceptor covalently bound in a distinct distance on the oligonucleotide strands were necessary. As a consequence, the whole spectrum of different methods for phosphoramidite syntheses and protocols for oligonucleotide modifications have been applied, developed, and improved dramatically during the last 10–15 years in order to prepare structurally well-defined DNA systems [9]. Significant experimental improvement came with the DNA assays bearing charge donors that have been attached covalently to oligonucleotides. Using these DNA systems, a systematic measurement of the distance dependence and the base sequence dependence of the charge transfer processes became possible.

Accordingly, most of the past studies of DNA-mediated charge transfer processes were performed with respect to the following strategies [10]:

1. Covalent labeling of the DNA with redox-active probes.
2. Photochemical initiation of the charge transfer process.
3. Spectroscopic or electrochemical detection of the charge transfer processes or analysis of irreversible DNA products yielded by the charge transfer reaction.

A broad variety of organic or inorganic intercalators, sugar modifications, and natural or modified DNA bases have been used as charge donors and charge acceptors in order to study charge transfer or transport phenomena in DNA. In principle, the existing DNA assays can be classified by their characteristic structural features (Figure 1.1) [9]:

A. DNA duplexes with unnatural or artificial DNA bases.
B. DNA duplexes with DNA base modifications pointing into the major or minor groove.
C. Capped DNA hairpins with a duplex stem.
D. DNA duplexes with organic or inorganic intercalators covalently attached via a flexible linker.

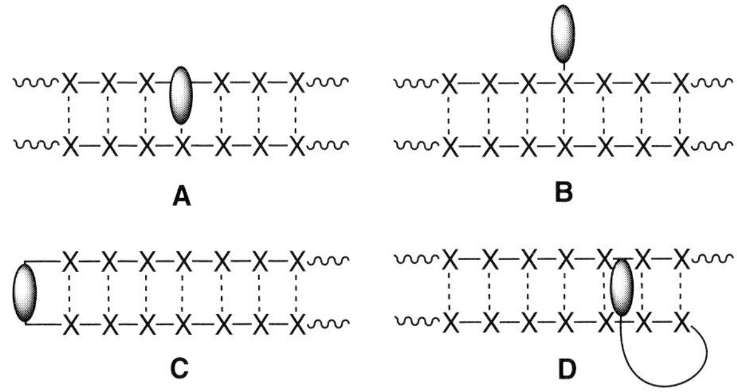

Fig. 1.1 Principal structures **A–D** of DNA donor-acceptor assays for the investigation of charge transfer processes in DNA.

In cases A, B, and C, the DNA base or sugar modifications can be introduced via automated solid-phase synthetic methods using suitable DNA building blocks. As an alternative route, DNA modifications can be introduced by solid-phase methods that are applied during or after the completed automatic solid-phase synthesis, as is the case for preparation of DNA assays B and D.

1.3
Photoinduced Oxidative Hole Transfer vs. Reductive Electron Transfer in DNA

In principle, DNA-mediated charge transfer processes can be categorized as either oxidative hole transfer or reductive electron transfer processes (Figure 1.2) [11, 12]. The phrase hole transfer is somewhat misleading since it describes an electron transfer in the opposite direction. Both processes, hole transfer and electron transfer, are in fact electron transfer reactions. However, with respect to orbital control it becomes clear that this categorization is not just a formalism about the different direction of the electron transfer. The oxidative hole transfer is a HOMO-controlled process, whereas the reductive electron transfer is LUMO-controlled.

With respect to biological motivation, most research groups have initially focused their work on the photochemically or photophysically induced oxidation of DNA and, furthermore, on the mobility of the created positively charged radical in the DNA. In this case, an electron is transferred from the DNA or the final acceptor (**A**) to the photoexcited charge donor (**D**). Such processes can be described as *oxidative hole transfer* or *hole transport*. On the other hand, charge transfer plays a growing role in the recent development of DNA-based nanowires and DNA microarrays detecting single-base mismatches or various DNA lesions by electrochemical readout methods [7]. For these purposes, the mobility of an excess electron in the DNA base stack has been applied. *Reductive electron transfer* or *excess electron transport* occurs if the photoexcited electron of **D** is injected into the DNA or transferred to the final electron acceptor **A**. In contrast to the broad and detailed knowledge about oxidative hole transfer and hole hopping in DNA, the mechanistic de-

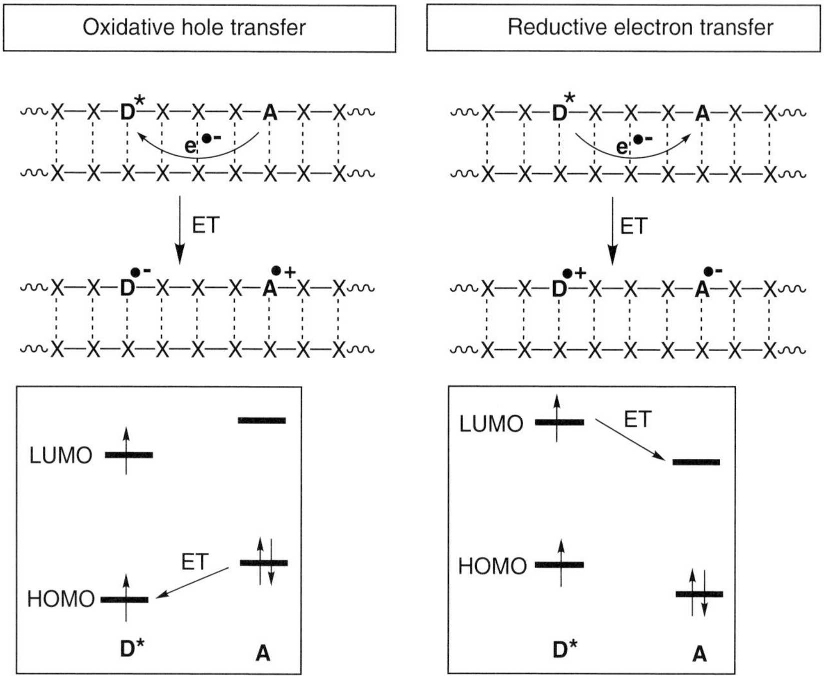

Fig. 1.2 Comparison of photoinduced oxidative hole transfer (HOMO control) and reductive electron transfer (LUMO control) in DNA (**D** = donor, **A** = acceptor, CT = charge transfer).

tails of excess electron transfer and migration remain unclear. Some lack of knowledge has been filled considerably during the last 2–3 years.

1.4
Hole Transfer and Hole Hopping in DNA

1.4.1
Spectroscopic Studies and Mechanisms of Hole Transfer in DNA

When the investigations of DNA-mediated charge transfer were started, most research groups interpreted their results according to the Marcus theory of nonadiabatic electron transfer [13]:

$$k_{\mathrm{ET}} = \frac{4\pi^2 |V_{\mathrm{el}}|^2}{h} \sqrt{\frac{1}{4\pi\lambda k_{\mathrm{B}}\mathrm{T}}} \exp\left(\frac{-\{\Delta G_{\mathrm{ET}} + \lambda\}^2}{4\lambda k_{\mathrm{B}}\mathrm{T}}\right) \quad (1)$$

Three important quantities affect the rate of the electron transfer process (k_{ET}): the electronic coupling (V_{el}), the driving force of the electron transfer process

(ΔG_{ET}), and the reorganization energy (λ). A full understanding of DNA-mediated charge transfer requires knowledge of how these three variables are affected by the medium DNA as the bridge between donor **D** and acceptor **A**. In most synthetic DNA-donor-acceptor systems, this is not completely clear. In fact, especially the β-value has been highly over-interpreted, which was a major cause of the controversy. It is important to point out that the energetic level of the DNA bridge in relation to the energetic levels of **D** and **A** determines molecular wire-like behavior or a charge transfer via the superexchange mechanism (Figure 1.3). In the case of a molecular wire, the bridge states are energetically comparable to the level of **D** and the electron can be injected into the bridge. Upon injection, the electron is localized within the bridge and moves incoherently to **A**. However, in the case of the superexchange mechanism, the bridge states lie above the level of **D**. Consequently, the electron is transferred in one coherent jump and is never localized within the bridge. For the superexchange mechanism, the distance dependence behavior of k_{ET} is clearly exponential.

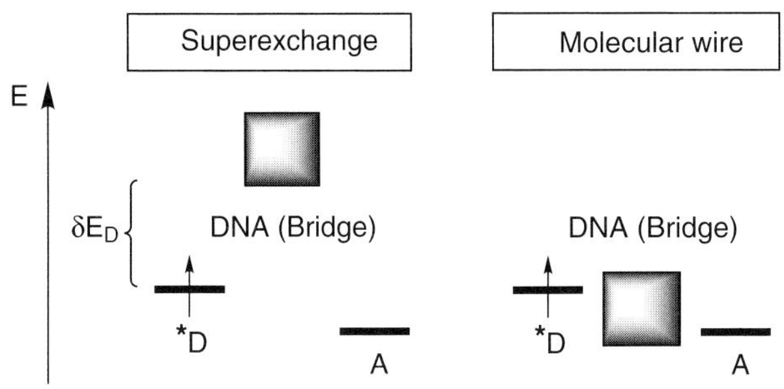

Fig. 1.3 Comparison of charge transfer via superexchange and via a molecular wire (**D** = donor, **A** = acceptor).

DNA represents a very special medium in terms of charge transfer processes. The planar heterocyclic aromatic systems of the DNA bases are stacked at a distance of 3.4 Å, which brought up the idea that DNA could conduct electrons like a molecular wire. Despite some initial controversies [2], it turned out that DNA as a medium for electron transfer is *not* a molecular wire. Accordingly, the DNA-mediated hole transfer processes were described in terms of a superexchange mechanism. The charge tunnels in one coherent step from **D** to **A** and never resides on the DNA bridge (**B**) between the two (Figure 1.4). This type of mechanism occurs if the bridge states are energetically higher than the photoexcited donor state (**D***). The rate k_{HT} of such a single-step process depends exponentially on the distance R between **D** and **A**, $k_{HT} \propto \exp(-\beta \cdot R)$. β represents the crucial parameter to describe the distance dependence of hole transfer in DNA which itself is dependent on the nature of the bridge **B** and its coupling with **D** and **A**. Values of β for charge transfer through proteins lie in the range of 1.0–1.4 $Å^{-1}$ [14]. In comparison, apparent β-values determined for hole transfer reactions in DNA can be found in a wide range from $\beta < 0.1$ $Å^{-1}$ to $\beta = 1.5$ $Å^{-1}$ (Table 1.1) [4].

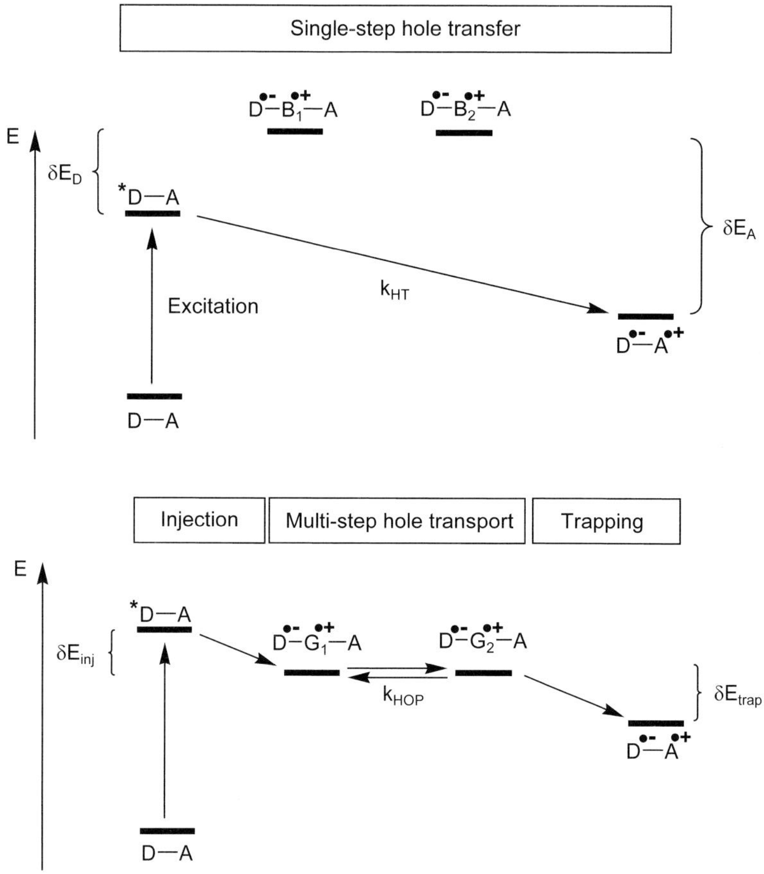

Fig. 1.4 Comparison of photoinduced DNA-mediated hole transfer via the superexchange mechanism and photoinduced hole transport via hopping (**D** = donor, **A** = acceptor, **B** = base, HT = hole transfer, inj = injection, HOP = hopping, trap = trapping).

Four important observations were drawn from this interpretation:

1. The hole transfer via the superexchange mechanism is limited to short distances (<10 Å).
2. Short-range hole transfer reactions occur on a very fast time scale ($k_{HT} = 10^9$–10^{12} s^{-1}).
3. The typical β-value of DNA-mediated hole transfer is 0.6–0.8 $Å^{-1}$.
4. The intercalation of the charge donor and acceptor is crucial for a fast and efficient hole transfer process.

The occurrence of very small β-values (≤ 0.1 $Å^{-1}$) exhibiting shallow distance dependence led to the description of an alternative mechanism, the hopping model (Figure 1.4) [5, 11]. Among the four different DNA bases, guanine (G) can

Table 1.1 Important spectroscopic studies of DNA-mediated hole transfer using DNA with covalently attached donor and acceptor.

Charge donor	Charge acceptor	β (Å^{-1})	k_{CT} (s^{-1})	Research group	Year
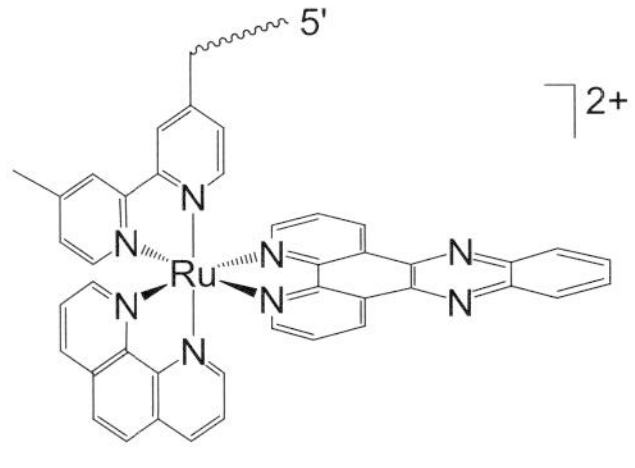	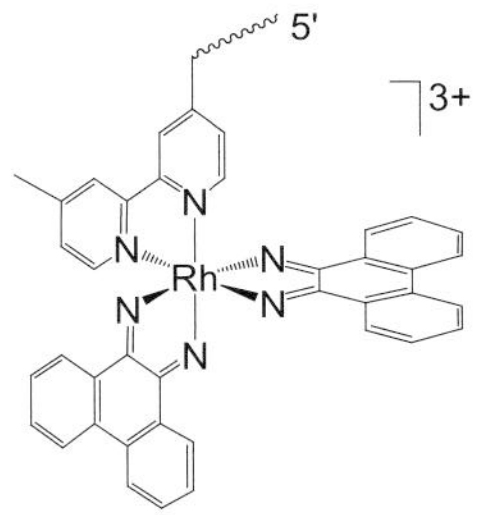	0.2	10^9	Barton et al. [3]	1993
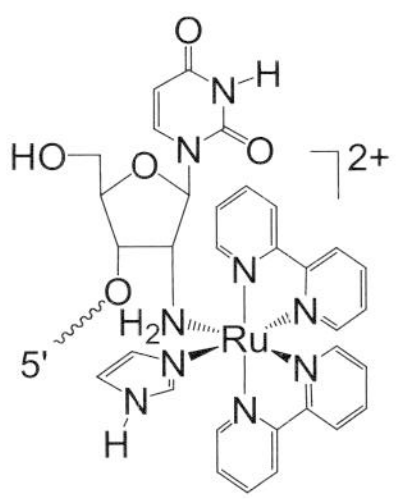	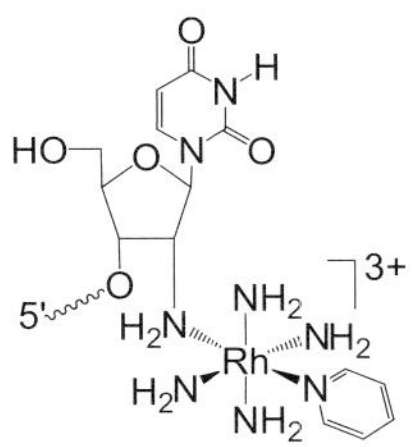	1.0–1.5	10^6	Meade et al. [15]	1995
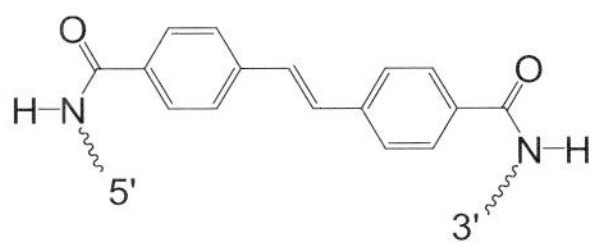	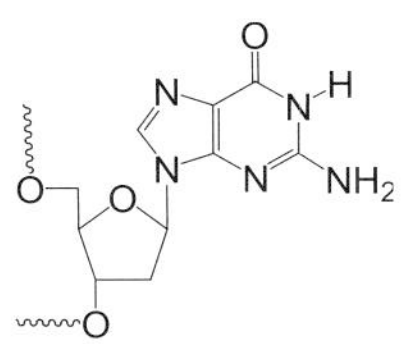	0.64	10^8–10^{12}	Lewis et al. [16]	1997

Table 1.1 (continued)

Charge donor	Charge acceptor	β (Å^{-1})	k_{CT} (s^{-1})	Research group	Year
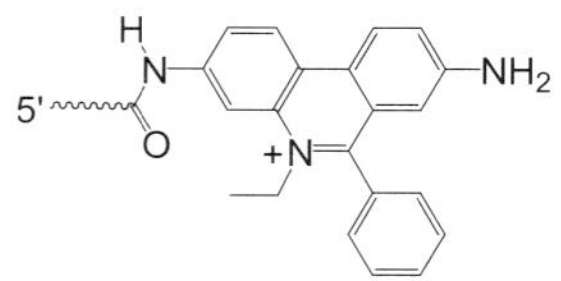	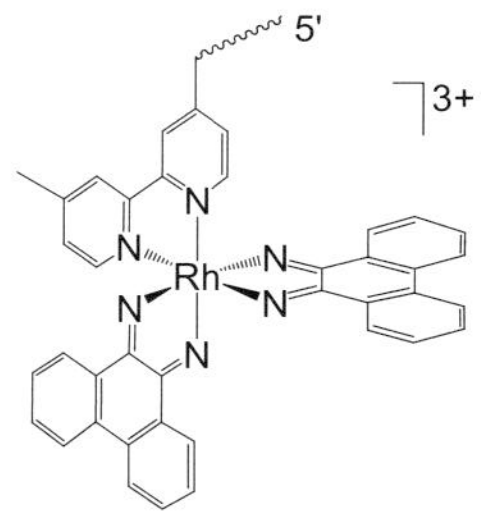	–	10^{10}	Barton et al. [17]	1997
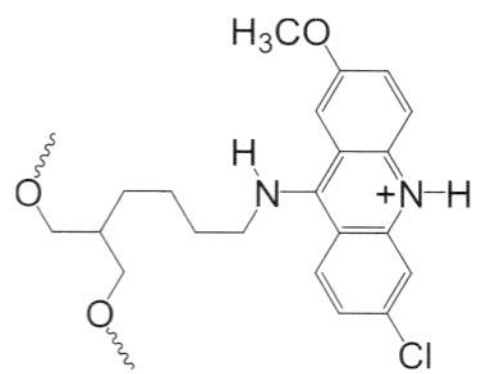	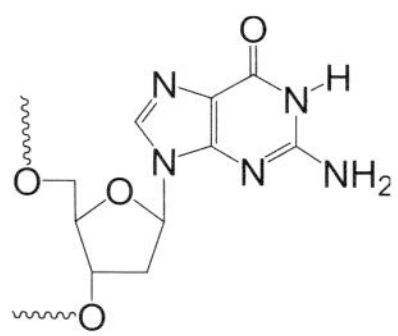	1.4	10^{5}–10^{10}	Tanaka et al. [18]	1998
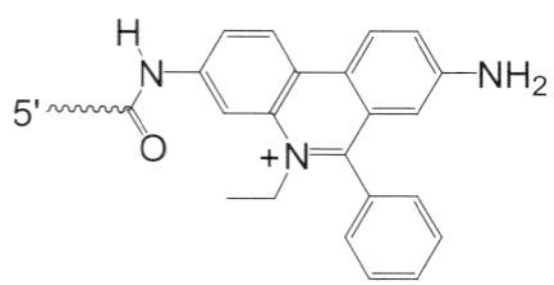	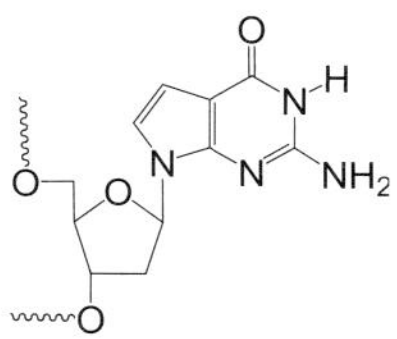	–	10^{12}	Barton/Zewail et al. [19]	1999

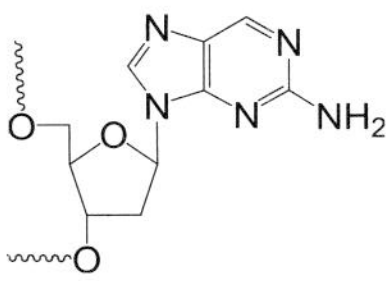	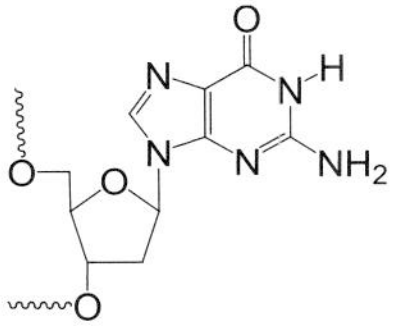		0.1–1.0	10^{9}–10^{11}	Barton/Zewail et al. [20]	1999
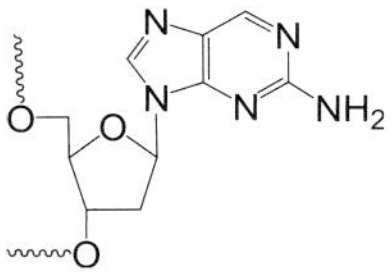	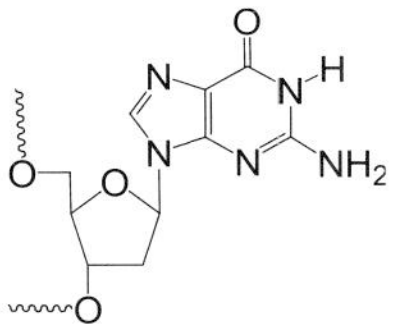		0.75	10^{6}–10^{7}	Shafirovich et al. [21]	2000
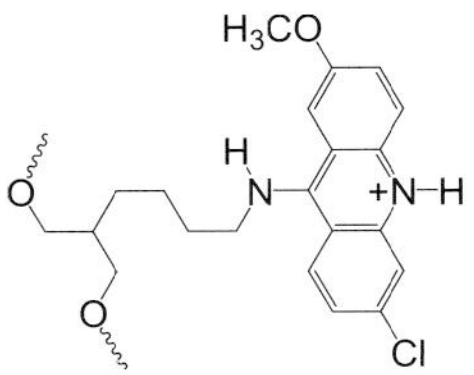	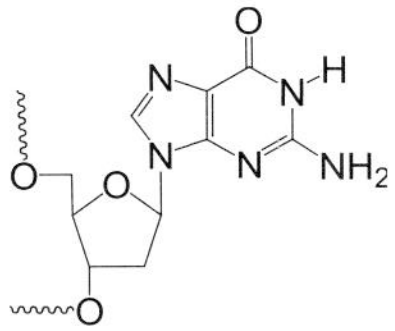	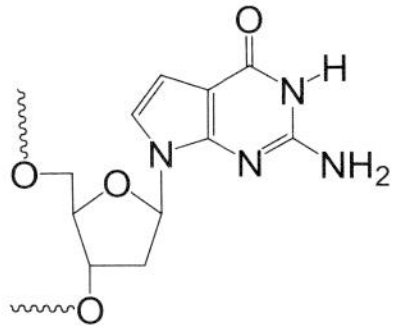	2.0/0.8	10^{7}–10^{12}	Michel-Beyerle et al. [22]	2001
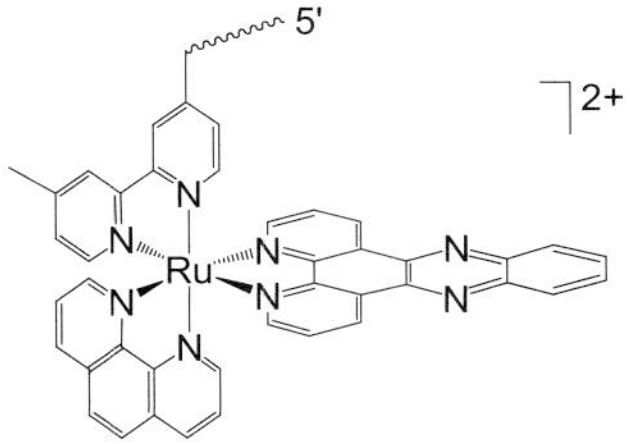	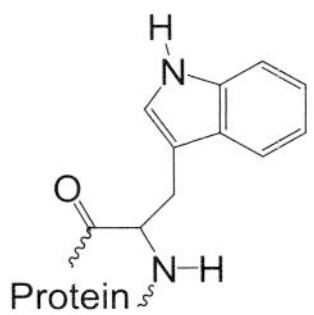	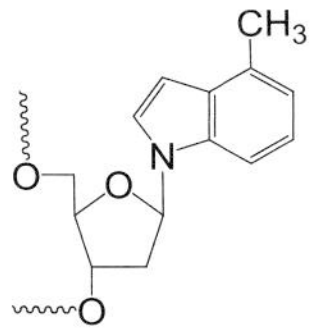	–	$\geq 10^{7}$	Barton et al. [23, 24]	2002

be most easily oxidized [25, 26]. Hence, the G radical cation plays the role of the intermediate charge carrier during the hopping process. In contrast to the previously described superexchange mechanism, the bridge and donor levels have to be similar in order to inject a hole thermally into the DNA base stack. Subsequently, the positive charge hops from G to G and can finally be trapped at a suitable charge acceptor. If each single hopping step occurs over the same distance, then the dynamics of hopping displays a shallow distance dependence with respect to the number of hopping steps N [5]:

$$k_{ET} = N^{-\eta} \tag{2}$$

The value of η lies between 1 and 2 and represents the influence of the medium. Each hopping step itself is a superexchange process through the intervening adenine-thymine (A-T) base pairs, but only if the A-T stretch is not too long (see below). The rate for a single hopping step from G to GG was determined to be $k_{HOP} = 10^6 - 10^8\ s^{-1}$ [27]. Using the site-specific binding of methyltransferase *HhaI* to DNA, a lower limit for hole hopping in DNA $k_{HT} > 10^6\ s^{-1}$ was measured over 50 Å through the base stack [24]. Based on the absence of a significant distance dependence, it was concluded that hole hopping through the DNA is not a rate-limiting step.

Recently, it was proposed and underscored with experimental evidence that adenines can play the role of intermediate hole carriers (Figure 1.5) [28]. Such A-hopping can occur if G is not present within the sequential context, mainly in longer A-T stretches (at least four A-T base pairs) between the guanines (Figure 1.5). The oxidation of A by $G^{\bullet +}$ is endothermic. With respect to the low efficiency of this hole-hopping step, it was suggested that once $A^{\bullet +}$ has been generated, the A-hopping proceeds fast. In fact, the rate of A-hopping has been determined to be $k_{HT} = 10^{10}\ s^{-1}$ [29]. Moreover, it could be shown that hole transport over eight A-T base pairs is nearly as efficient as the hole transport over two A-T base pairs [30]. In comparison to G-hopping, A-hopping proceeds faster, more efficiently, and almost distance independently. Recent calculations suggest that hole transport through stacked A-T base pairs may be most favorable [31]. It is known from γ-radiolysis studies that the one-electron oxidation of DNA bases has drastic effects on their acidity. In theory, proton transfer processes could occur on time scales comparable to charge transfer reactions and can therefore dramatically influence the charge transport efficiency due to the separation of spin and charge [32]. The question of proton transfer in oxidized $G^{\bullet +}$-C base pairs is crucial for the understanding of hole hopping in DNA. The pK_a value of $G^{\bullet +}$ is ~3.9 [33]. The pK_a value of the complementary DNA base cytosine (C) is very similar (4.5) [33]. Hence, there is likely an equally distributed protonation equilibrium in a one-electron oxidized $G^{\bullet +}$-C base pair that is principally reversible but could interfere with the hole transport and potentially interrupt hole hopping in DNA. In fact, measurements of the kinetic isotope effect of hole transport in DNA have been performed and provide some evidence for a coupling between hole hopping and proton transfer processes [34]. The situation is different in the A-T pair. Oxidized adenine ($A^{\bullet +}$) represents a

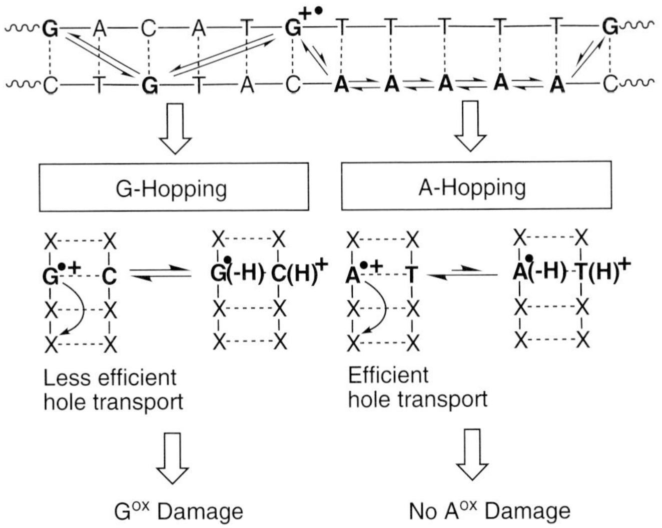

Fig. 1.5 DNA-mediated hole transport via guanine hopping and adenine hopping.

powerful acid with a pK_a of 1, and T shows an extremely low basicity (pK_a of $T(H)^+$ is –5) [33]. Taken together it becomes clear that charge and spin remain located on the A in an $A^{\bullet+}$-T base pair (Figure 1.5).

The discovery of A-hopping is an excellent example of how the mechanistic picture about hole transport and hole transfer processes through DNA becomes more complex and simultaneously more complete. For instance, a polaron-like model was suggested as a more precise mechanistic description for hole transport through DNA [35]. The polaron represents a structural distortion of the DNA that stabilizes and delocalizes the radical cation over several DNA bases. On the other hand, a formal relation between the charge transfer rate through a molecular bridge and the conductance of this bridge has been established [36]. This means that dephasing and relaxation effects influence the charge transfer rate. It is clear that the role of base motions in delocalization and propagation of charges through the DNA duplex must be taken into account. Most recently, an oscillatory component of the distance dependence with a period of 4–5 base pairs has been described experimentally and cannot be rationalized by the phonon-assisted polaron-hopping model or by a temperature-induced transition from superexchange to thermally induced hopping [37]. As a result, hole transport through DNA has been described as a conformationally gated hopping among stacked domains, the so-called charge transfer-active domains representing a conformation of a few well-coupled DNA bases that exists only in a distinct time frame [37].

1.4.2
Biochemical and Chemical Hole Trapping in DNA

Besides A-hopping, $G^{\bullet+}$ plays the major role of the intermediate charge carrier during the hole-hopping process. The G radical cation was identified as the precursor of a variety of different oxidative G lesions, which are normally described as G^{ox}. Some of these G oxidation products have been identified [6]. Bioanalytical experiments explore hole transport reactions through DNA by an indirect method. As described above, after the photochemical or photophysical oxidation of DNA using a suitable intercalator, G is preferentially oxidized. The resulting $G^{\bullet+}$ can react with H_2O and/or O_2, yielding G^{ox}. Such modified DNA strands can be cleaved at the site of G^{ox} by treatment with, e.g., piperidine at elevated temperature and then separated by gel electrophoresis and visualized by phosphorimagery using radioactive ^{32}P-labeling.

The most common photooxidants for DNA are metal complexes or organic intercalators such as Rh(III) complexes, Ru(II) complexes, ethidium derivatives, anthraquinone derivatives, uridine modified with cyanobenzoquinones, and modified 2′-deoxyribosides bearing a photoreactive group (Figure 1.6) [10]. These systems differ significantly in their structural properties, their redox potentials, and their absorbing wavelengths. It has been observed in all systems that the positive charge can be transported with high efficiency over very long distances (up to 200 Å) [38]. The observed efficiency of hole transport seems to be strongly dependent on the integrity of the conformation of the intervening DNA base pairs.

The hole injection system by Giese et al. represents an important exception in comparison to the others since it relies on a Norrish type I cleavage reaction of a 4'-modified uridine derivative yielding a sugar radical cation (Figure 1.7 a) [39]. This enol ether radical cation exhibits a higher oxidation potential compared to G and hence is able to inject a hole into the nearest G within the DNA duplex. This hole injection system works from the ground state, although it is photoinitiated, and thus has the advantage that fast back charge transfer processes do not occur. Additionally, this assay allows elucidation of some kinetic information about hole hopping in DNA, since the trapping reaction of the enol ether radical cation by water competes with the hole injection process and exhibits a rate of 10^8 s^{-1}.

The oxidation of G yielding the piperidine-labile G^{ox} represents probably the most prominent example of a thermodynamic hole trap for hole hopping in DNA. Additionally, chemically modified DNA bases have been presented as new and interesting kinetic hole traps that allow the site-specific trapping of holes in DNA and the chemical probing of hole hopping. Especially cyclopropyl-modified guanine (Cp-G) [40] and adenine (Cp-A) [41] have been applied, since the ring opening of the *N*-alkylcyclopropylaminyl radical [42] can be considered as a radical clock (Figure 1.7 b). Cp-A in particular represents an important tool for the elucidation of adenine hopping when it is incorporated into longer A-T bridges between two GG sites. As mentioned previously in this chapter, oxidative adenine damage (A^{ox}) has not yet been observed as a result of hole hopping over longer A-T bridges and involving $A^{\bullet+}$ as an intermediate charge carrier. In fact, using

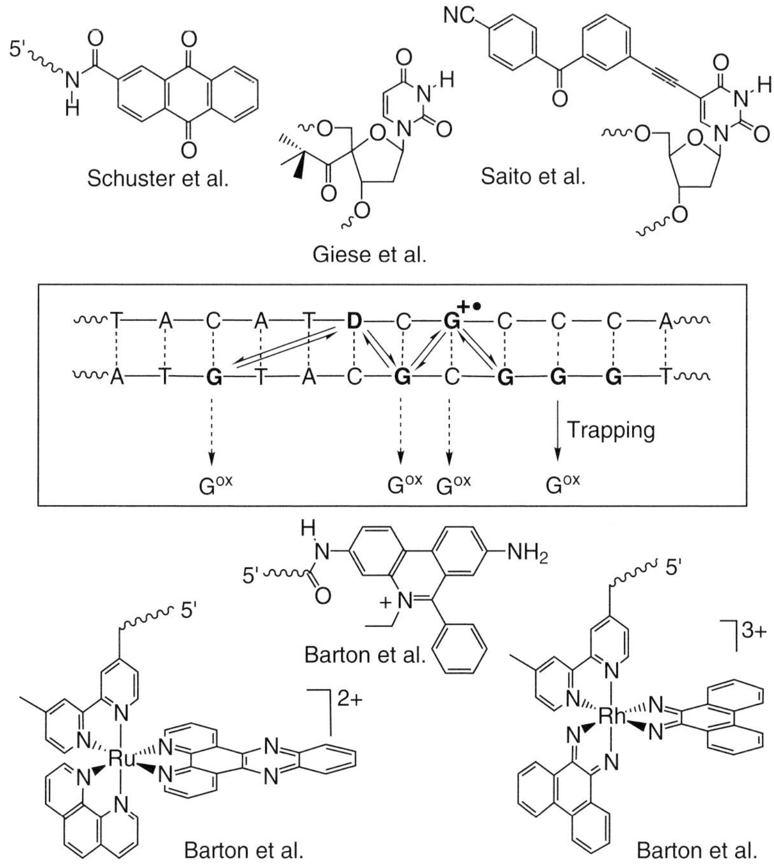

Fig. 1.6 Important examples of photooxidants (**D**) for the investigation of DNA-mediated hole transport by bioanalytical detection of oxidative guanine damage (G^{ox}).

Cp-A, the hole could be trapped within the A-T bridge, which proves the presence of a positive charge in this area [41]. Hence, cyclopropyl-substituted nucleosides are promising tools to prove the existence of the transient radical species in DNA. The corresponding chemical trapping is in agreement with the previous consideration that nucleobases possessing higher oxidation potentials than G, such as A, are also able to participate directly in the multi-step hopping mechanism.

1.4.3
Modulation of DNA-mediated Hole Transfer

DNA-mediated hole transfer exhibits an extremely high sensitivity to the π-stacking of the intervening DNA bases as well as to disruption and perturbation of the DNA structure or conformation. Hence, checking the DNA integrity and sensing

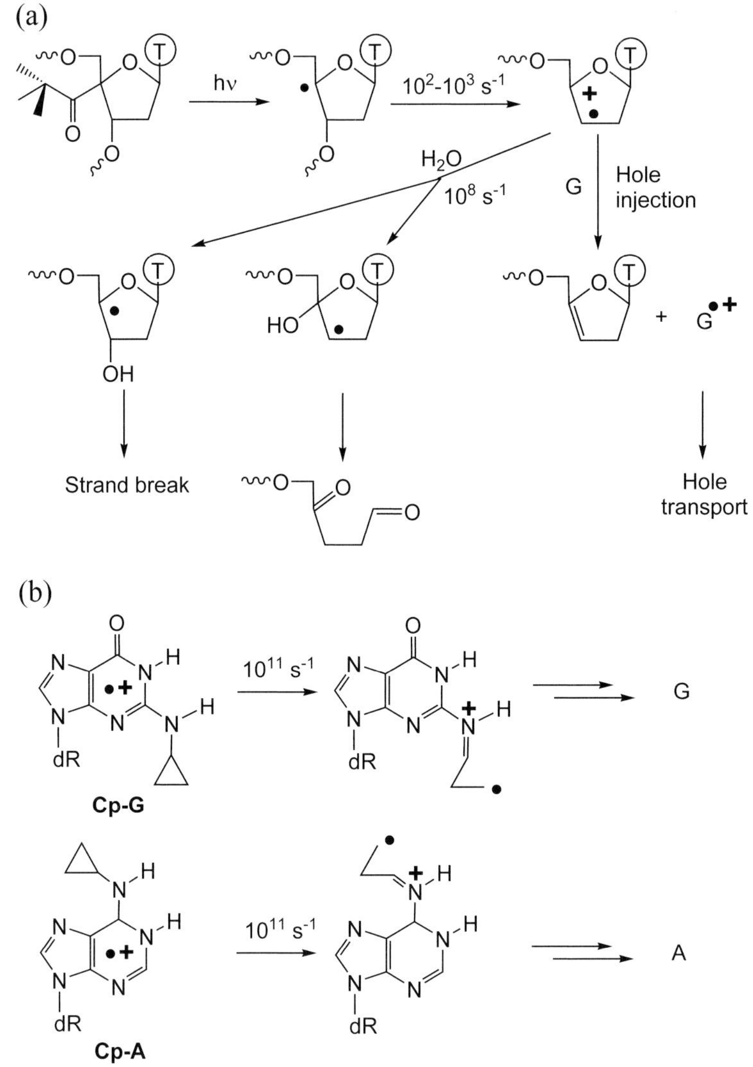

Fig. 1.7 Chemical hole-injection (a) and hole-trapping (b) assays.

of DNA damage at the various checkpoints during the cell cycle could be accomplished based upon charge transfer reactions. Recently, charge transfer in HeLa cell nuclei has been probed using a rhodium photooxidant [43]. After incubation and irradiation, the genomic DNA was isolated and analyzed, revealing that the base damage occurs preferentially at the 5′-G of GG sites. More importantly, oxidative G damage was found at protein-bound sites that were inaccessible to the rhodium photooxidant as examined by footprinting. This clearly indicates that hole transfer processes can occur in cells.

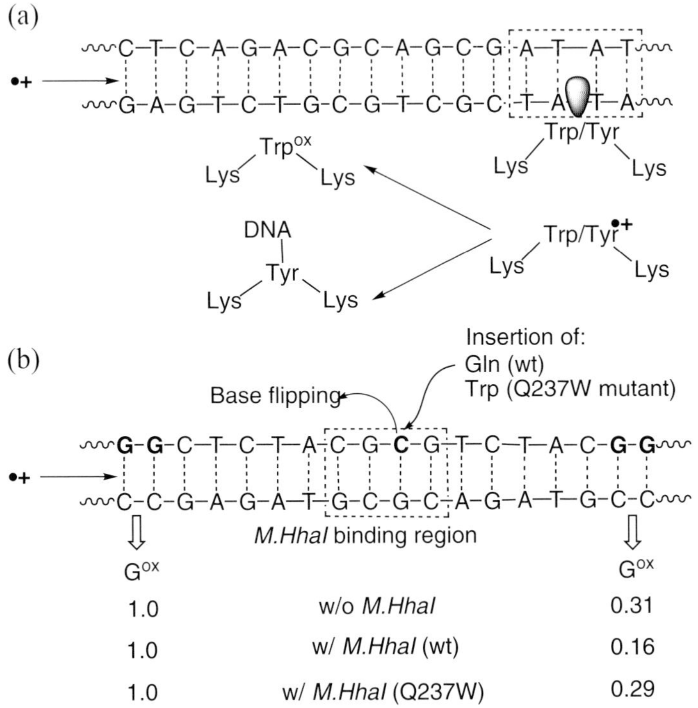

Fig. 1.8 (a) DNA-to-peptide hole transport. (b) Modulation of DNA-mediated hole transport by cytosine methyl transferase, *HhaI* wild-type, and Gln237 Trp mutant.

It is very reasonable to begin investigation of protein-modulated hole transport in DNA by using DNA-binding peptides. In principal, aromatic amino acids like tyrosine (Tyr) or tryptophan (Trp) have a lower oxidation potential than G and can therefore be used as peptidic traps for holes [44]. The smallest-possible DNA-binding peptides Lys-Trp-Lys and Lys-Tyr-Lys have been applied, and it was shown that the radicals of Trp and Tyr can be generated by DNA-mediated hole transport and can occur on the microsecond time scale in high yields (Figure 1.8 a) [45]. Hence, DNA-binding peptides as models for proteins play an important role in protecting the genome from radical damage. Interestingly, the generated peptide radicals exhibit a completely different reactivity. The DNA-bound Trp radical forms oxidized products in the presence of O_2, while the DNA-bound Tyr radical forms cross-links with the DNA bases at the peptide-binding site.

Specific DNA-protein interactions that either promote or inhibit hole transfer processes would be the most crucial part in biological charge transfer systems. The following experiments have shown clearly that DNA-mediated hole transfer processes are modulated both negatively and positively by DNA-binding proteins. Most importantly, each of the observed influences of the proteins can be explained by special structural features of the corresponding DNA-protein complexes.

Hence, specific DNA-protein interactions result in a characteristic modulation of the DNA-mediated hole transport.

One of the most interesting examples of how proteins can modulate charge transfer in DNA is the cytosine methyltransferase *HhaI* [23, 46]. This protein recognizes the sequence 5′-GCGC-3′, and during methylation a base-flipped complex is formed with the target C in an extrahelical position (Figure 1.8 b). A glutamine (Gln) side chain fills the space in the DNA duplex. Not surprisingly, the hole transport in the DNA was attenuated as a result of the interruption of the base stack. The situation is different when using an M.*HhaI* mutant containing Trp rather than Gln237 (Q237W). Interestingly, long-range hole transport was restored upon binding with this Trp mutant of M.*HhaI*. This is the result of an electronic interaction of the flat aromatic indole heterocycle of Trp with the neighboring DNA bases. In conclusion, the indole side chain of Trp is able to replace a normal DNA base by electronic means [46]. The M.*HhaI*-DNA complex using the Q237W mutant was applied to time-resolved transient absorption spectroscopy [23]. The product radical was identified as a mixture of the Trp and G radicals occurring in the DNA-protein contact area. By laser experiments it was possible to establish a lower limit for hole transport in DNA of $k_{HT} > 10^6\ s^{-1}$ through 50 Å through the DNA base stack. Based on the absence of significant distance dependence, hole transport through the DNA is not rate limiting.

In contrast to wild-type M.*HhaI*, DNA-protein interactions could facilitate hole transport in DNA. Especially proteins that bind to the major groove but do not perturb the normal B-DNA structure can significantly enhance hole transport efficiency in DNA. This was demonstrated by using either the restriction endonuclease *PvuII* or the transcription factor ANTP (Figure 1.9) [47]. As a result of the binding of proteins, the DNA conformation is stiffened, the conformational movements are diminished, and, consequently, the hole transport is facilitated. In contrast to R.*PvuII* and ANTP, the TATA box-binding protein induces two 90-degree bends into the DNA duplex. Due to this strong conformational change in the helix structure, the hole transport efficiency decreases significantly [47]. More recently, the complex of endonuclease *BamHI* and the target DNA was investigated by guanine oxidation as the result of a DNA-mediated hole transport [48]. In this case, the direct contact of a positively charged guanidinium group of the protein to the recognition sequence of the DNA completely suppressed hole transport and dramatically lowered the guanine damage efficiency (Figure 1.9).

The counterions also play an important role in DNA-mediated hole transport. Interestingly, Schuster et al. elucidated that the migration of charges in DNA can be gated by ions [49]. However, a strong dependence of the efficiency of hole transport on the identity of the counterions is unlikely. Most recently, there have been attempts to modulate hole transport through DNA by artificial DNA bases that can tune the π-stacking properties within the DNA duplex. For instance, benzo-fused adenine bears a larger aromatic surface and enhanced stacking properties, thereby providing better hole transfer ability [50].

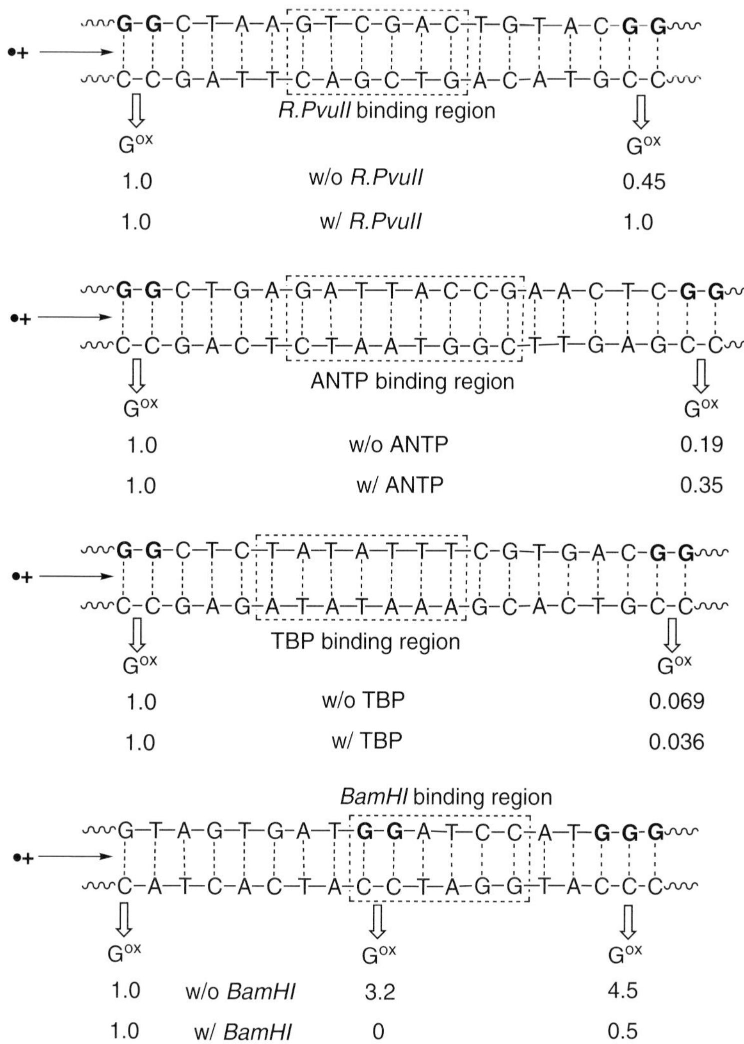

Fig. 1.9 Modulation of DNA-mediated hole transport by the following proteins: restriction endonucleases *PvuII* and *BamH1*, antennapedia homeodomain protein (ANTP), and TATA box-binding protein (TBP).

1.5 Reductive Electron Transfer in DNA

1.5.1 Mechanisms of Electron Transfer in DNA

In contrast to the broad knowledge available about oxidative hole transfer and hole hopping, as described in the previous sections, the mechanistic details of ex-

are immobilized as single-strand oligonucleotides on an electrode or chip and contain a redox-active probe that is intercalated and/or covalently attached (Figure 1.13). Intact DNA material added to the chip forms intact DNA duplexes leading to an efficient electron transfer between the chip surface and the distant redox-active probe. Base mismatches and DNA lesions significantly interrupt charge transfer in DNA, and as a result, the electrochemical response is missing.

One of the most convenient techniques for the depositing of biopolymers on solid-phase surfaces is the self-assembled monolayer [65]. According to this technique, DNA is attached to an alkyl thiolate linker via the 5′-terminal hydroxy group of the oligonucleotide, which then interacts with the gold electrode to form DNA films. Additionally, the DNA is labeled with redox-active probes, such as daunomycin, pyrrolo-quinoline-quinone, methylene blue, or ferrocene [7]. Using this methodology, a broad range of single-point mutations and DNA lesions can be detected without the context of certain base sequences. Hence, electron transfer through DNA films offers a new and suitable approach for the development of sensitive DNA sensors and chips. Normally, sensitive gene detection is accomplished by the amplification of the DNA material through the polymerase chain reaction (PCR). Inherent limitations of PCR often prohibit this application. Thus, research in the field of new DNA chips is currently focused on increasing sensitivity in such a way that PCR amplification becomes unnecessary.

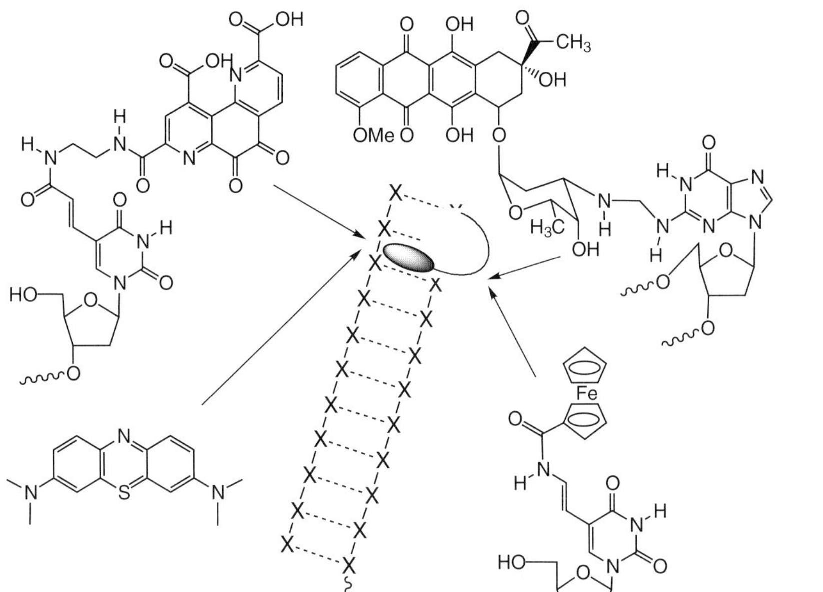

Fig. 1.13 Examples of redox-active probes for an electrochemical readout on DNA chips and microarrays.

1.6 Conclusions

This introductory chapter summarizes briefly all of the important and basic aspects related to charge transfer processes in DNA. This information should underscore (in addition to a variety of detailed questions) that a pretty clear picture about the phenomenon of "charge transfer processes in DNA" has emerged by now. The extreme controversy has been solved by very differential interpretations of the applied DNA systems and the description of alternative mechanisms.

In conclusion, it has turned out that excess electron transport occurs via a hopping mechanism over long distances (Figure 1.12), which is comparable to a certain extent with hole hopping in DNA (Figure 1.5). But it is important to note here that there are significant differences between both types of charge transport. First, hole transport occurs preferentially via guanine hopping and, only in stretches longer than 3–4 A-T base pairs, via adenine hopping. Hence, the hole-hopping process can be divided into distinct sequence regimes. In contrast, the electron transport occurs via mixed cytosine and thymine hopping, with some preference for the thymine radical anion as the intermediate electron carrier. Second, during hole hopping, the irreversible oxidation of the guanine radical cation yielding the G^{ox} damage competes with the hole transport. In contrast, no damage as a result of excess electron migration has been detected yet. This means that hole transport can never occur without causing damage, whereas electron transport potentially can.

The latter conclusion has important significance for the biotechnical application of DNA-mediated charge transfer as well as for the proposed biological role of charge transfer during DNA damage recognition. In both cases it can be assumed that it is better to use the transport of an excess electron rather than a hole during the sensing procedure in order to avoid DNA damage.

Despite this broad knowledge, DNA research is still far from a profound and clear understanding of the electronic properties and interactions in DNA. In the past, DNA-mediated charge transfer has been the subject of considerable interest, having biological relevance in the formation and repair of lesions and damage in DNA. The future will show the high potential for applications of these processes in the development of new DNA assays and microarrays as well as DNA-inspired devices for nanotechnology.

References

1 D. D. Eley, D. I. Spivey, *Trans. Faraday Soc.* **1962**, *58*, 411–415.

2 a) S. Priyadarshy, S. M. Risser, D. N. Beratan, *J. Phys. Chem.* **1996**, *100*, 17678–17682; b) N. J. Turro, J. K. Barton, *J. Biol. Inorg. Chem.* **1998**, *3*, 201–109; c) Y. A. Berlin, A. L. Burin, M. A. Ratner, *Superlattices Microstruct.* **2000**, *28*, 241–252.

3 C. J. Murphy, M. R. Arkin, Y. Jenkins, N. D. Ghatlia, S. H. Bossmann, N. J. Turro, J. K. Barton, *Science* **1993**, *262*, 1025–1029.

4 See special issues of *Top. Curr. Chem.* **2004**, *236* and *237*; and references therein.

5 J. Jortner, M. Bixon, T. Langenbacher, M. E. Michel-Beyerle, *Proc. Natl. Acad. Sci. USA* **1998**, *95*, 12759–12765.

6 a) P. O'Neill, E. M. Frieden, *Adv. Radiat. Biol.* **1993**, *17*, 53–120; c) C. J. Burrows, J. G. Muller, *Chem. Rev.* **1998**, *98*, 1109–1151; c) D. Wang, D. A. Kreutzer, J. M. Essigmann, *Mutation Res.* **1998**, *400*, 99–115; e) S. Kawanashi, Y. Hiraku, S. Oikawa, *Mutation Res.* **2001**, *488*, 65–76.

7 See review and references therein: T. G. Drummond, M. G. Hill, J. K. Barton, *Nature Biotechnol.* **2003**, *21*, 1192–1199.

8 See review: D. Porath, G. Cuniberti, R. Di Felice, *Top. Curr. Chem.* **2004**, *237*, 183–227.

9 See review and references therein: H.-A. Wagenknecht, *Curr. Org. Chem.* **2004**, *8*, 251–266.

10 See review and references therein: M. W. Grinstaff, *Angew. Chem. Int. Ed.* **1999**, *38*, 3629–3635.

11 See review: B. Giese, *Annu. Rev. Biochem.* **2002**, *71*, 51–70.

12 See review: H.-A. Wagenknecht, *Angew. Chem. Int. Ed.* **2003**, *42*, 2454–2460.

13 R. A. Marcus, N. Sutin, *Biochim. Biophys. Acta* **1985**, *811*, 265–322.

14 a) J. R. Winkler, H. B. Gray, *Chem. Rev.* **1992**, *92*, 369–379; b) M. R. Wasielewski, *Chem. Rev.* **1992**, *92*, 435–461.

15 T. J. Meade, J. F. Kayyem, *Angew. Chem. Int. Ed.* **1995**, *34*, 352–354.

16 F. D. Lewis, R. Wu, Y. Zhang, R. L. Letsinger, S. R. Greenfield, M. R. Wasielewski, *Science* **1997**, *277*, 673–676.

17 S. O. Kelley, R. E. Holmlin, E. D. A. Stemp, J. K. Barton, *J. Am. Chem. Soc.* **1997**, *119*, 9861–9870.

18 K. Fukui, K. Tanaka, K. *Angew. Chem. Int. Ed.* **1998**, *37*, 158–161.

19 C. Wan, T. Fiebig, S. O. Kelley, C. R. Treadway, J. K. Barton, A. H. Zewail, *Proc. Natl. Acad. Sci. USA* **1999**, *96*, 6014–6019.

20 a) S. O. Kelley, J. K. Barton, J. K. *Science* **1999**, *283*, 375–381; b) C. Wan, T. Fiebig, O. Schiemann, J. K. Barton, A. H. Zewail, *Proc. Natl. Acad. Sci. USA* **2000**, *97*, 14052–14055.

21 a) V. Shafirovich, A. Dourandin, W. Huang, N. P. Luneva, N. E. Geacintov, *Phys. Chem. Chem. Phys.* **2000**, *2*, 4399–4408.

22 S. Hess, M. Götz, W. B. Davis, M. E. Michel-Beyerle, M. E. *J. Am. Chem. Soc.* **2001**, *123*, 10046–10055.

23 H.-A. Wagenknecht, S. R. Rajski, M. Pascaly, E. D. A. Stemp, J. K. Barton, *J. Am. Chem. Soc.* **2001**, *123*, 4400.

24 M. Pascaly, J. Yoo, J. K. Barton, J. K. *J. Am. Chem. Soc.* **2002**, *124*, 9083–9092.

25 S. Steenken, S. V. Jovanovic, *J. Am. Chem. Soc.* **1997**, *119*, 617–618.

26 C. A. M. Seidel, A. Schulz, M. H. M. Sauer, *J. Phys. Chem.* **1996**, *100*, 5541–5553.

27 F. D. Lewis, X. Liu, J. Liu, S. E. Miller, R. T. Hayes, M. R. Wasielewski, *Nature* **2000**, *406*, 51–53.

28 B. Giese, M. Spichty, *ChemPhysChem* **2000**, *1*, 195–198; b) B. Giese, J. Amaudrut, A.-K. Köhler, M. Spormann, S. Wessely, *Nature* **2001**, *412*, 318–320.
29 T. Takada, K. Kawai, X. Cai, A. Sugimoto, M. Fujitsuka, T. Majima, *J. Am. Chem. Soc.* **2004**, *126*, 1125.
30 B. Giese, T. Kendrick, *Chem. Commun.* **2002**, 2016–2017.
31 X. Li, Z. Cai, M. D. Sevilla, *J. Phys. Chem. A* **2002**, *106*, 9345–9351.
32 S. Steenken, *Biol. Chem.* **1997**, *378*, 1293–1297.
33 S. Steenken, *Free Rad. Res. Comms.* **1992**, *16*, 349–379.
34 B. Giese, S. Wessely, *Chem. Commun.* **2001**, 2108–2109.
35 G. B. Schuster, *Top. Curr. Chem.* **2004**, *236*, 139–161.
36 D. Segal, A. Nitzan, W. B. Davis, M. R. Wasielewski, M. A. Ratner, *J. Phys. Chem. B* **2000**, *104*, 3817–3829.
37 M. O'Neill, J. K. Barton, *J. Am. Chem. Soc.* **2004**, *126*, 11471–11483.
38 a) M. E. Núñez, D. B. Hall, J. K. Barton, *Chem. Biol.* **1999**, *6*, 85–97; b) P. T. Henderson, D. Jones, G. Hampikian, Y. Kan, G. B. Schuster, *Proc. Natl. Acad. Sci. USA* **1999**, *96*, 8353–8358.
39 See review: B. Giese, *Top. Curr. Chem.* **2004**, *236*, 27–44.
40 a) K. Nakatani, C. Dohno, I. Saito, *J. Am. Chem. Soc.* **2001**, *123*, 9681–9682; b) M. A. O'Neill, C. Dohno, J. K. Barton, *J. Am. Chem. Soc.* **2004**, *126*, 1316–1317.
41 C. Dohno, A. Ogawa, K. Nakatani, I. Saito, *J. Am. Chem. Soc.* **2003**, *125*, 10154–10155.
42 O. M. Musa, J. H. Horner, H. Shahin, M. Newcomb, *J. Am. Chem. Soc.* **1996**, *118*, 3862–3868.
43 M. E. Nunez, G. P. Holmquist, J. K. Barton, *Biochemistry* **2001**, *40*, 12465–12471.
44 See review: E. M. Boon, J. K. Barton, *Curr. Opin. Struct. Biol.* **2002**, 12, 320–329.
45 H.-A. Wagenknecht, E. D. A. Stemp, J. K. Barton, *J. Am. Chem. Soc.* **2000**, *122*, 1–7; b) H.-A. Wagenknecht, E. D. A. Stemp, J. K. Barton, *Biochemistry* **2000**, *39*, 5483–5491.
46 S. R. Rajski, S. Kumar, R. J. Roberts, J. K. Barton, *J. Am. Chem. Soc.* **1999**, *121*, 5615–5616.
47 S. R. Rajski, J. K. Barton, *Biochemistry* **2001**, *40*, 5556–5564.
48 K. Nakatani, C. Dohno, A. Ogawa, I. Saito, *Chem. Biol.* **2002**, *9*, 361–366.
49 R. N. Barnett, C. L. Cleveland, A. Joy, U. Landman, G. B. Schuster, *Science* **2001**, *294*, 567–571.
50 A. Okamoto, K. Tanaka, I. Saito, *J. Am. Chem. Soc.* **2003**, *125*, 5066–5071.
51 S. Steenken, J. P. Telo, H. M. Novais, L. P. Candeias, *J. Am. Chem. Soc.* **1992**, *114*, 4701–4709.
52 A. A. Voityuk, M.-E. Michel-Beyerle, N. Rösch, *Chem. Phys. Lett.* **2001**, *342*, 231–238.
53 See review: Z. Cai, M. D. Sevilla, *Top. Curr. Chem.* **2004**, *237*, 103–128.

54 a) C. Behrens, L. T. Burgdorf, A. Schwögler, T. Carell, *Angew. Chem. Int. Ed.* **2002**, *41*, 1763–1766; b) C. Haas, K. Kräling, M. Cichon, N. Rahe, T. Carell, *Angew. Chem. Int. Ed.* **2004**, *43*, 1842–1844.
55 a) T. Ito, S. E. Rokita, *J. Am. Chem. Soc.* **2003**, *125*, 11480–11481; b) T. Ito, S. E. Rokita, *Angew. Chem, Int. Ed.* **2004**, *43*, 1839–1842.
56 F. D. Lewis, X. Liu, S. E. Miller, R. T. Hayes, M. R. Wasielewski, *J. Am. Chem. Soc.* **2002**, *124*, 11280–11281.
57 C. Wagner, H.-A. Wagenknecht, *Chem. Eur. J.* **2005**, in press.
58 a) N. Amann, E. Pandurski, T. Fiebig, H.-A. Wagenknecht, *Chem. Eur. J.* **2002**, *8*, 4877–4883; b) P. Kaden, E. Mayer, A. Trifonov, T. Fiebig, H.-A. Wagenknecht, *Angew. Chem. Int. Ed.* **2005**, in press.
59 E. Rivera, R. H. Schuler, *J. Am. Chem. Soc.* **1983**, *87*, 3966–3971.
60 S.-R. Yeh, D. E. Falvey, *J. Am. Chem. Soc.* **1997**, *113*, 8557–8558.
61 B. Giese, B. Carl, T. Carl, T. Carell, C. Behrens, U. Hennecke, O. Schiemann, E. Feresin, *Angew. Chem, Int. Ed.* **2004**, *43*, 1848–1851.
62 a) R. Huber, T. Fiebig, H.-A. Wagenknecht, *Chem. Commun.* **2003**, 1878–1879; b) M. Raytchev, E. Mayer, N. Amann, H.-A. Wagenknecht, T. Fiebig, *ChemPhysChem* **2004**, *5*, 706–712.
63 See reviews and references therein: a) C. M. Niemeyer, D. Blohm, *Angew. Chem. Int. Ed.* **1999**, *38*, 2865–2869; b) D. H. Blohm, A. Guiseppi-Elie, *Curr. Opin. Biotechnol.* **2001**, *12*, 41–47; c) M. C. Pirrung, *Angew. Chem. Int. Ed.* **2002**, *41*, 1276–1289; d) A. Jung, *Anal. Bioanal. Chem.* **2002**, *372*, 41–42.
64 See special issue of *Nature Genetics* **1999**, *21*, 1–60; and references therein.
65 See reviews and references therein: a) A. Kumar, N. L. Abott, E. Kim, H. A. Biebuyck, G. M. Whitesides, *Acc. Chem. Res.* **1995**, *28*, 219–226; b) G. E. Porier, *Chem. Rev.* **1997**, *97*, 1117–1127.

2
Sequence-dependent DNA Dynamics: The Regulator of DNA-mediated Charge Transport

Melanie A. O'Neill and Jacqueline K. Barton

2.1
Introduction

More than a decade of clever experiments have shown unambiguously that DNA, the molecule responsible for the safekeeping of our genetic information, is anything but inert when it comes to the transport of charge [1]. Instead, double-helical DNA can mediate the efficient transport of charge, both electrons [2] and holes, on time scales as short as picoseconds [3–5]. Moreover, charge transport (CT) through DNA occurs over large molecular distances, at least as far as 200 Å [6, 7], leading to long-range oxidative damage [8] and repair [9]. We have exploited DNA's ability to transport charge, especially its sensitivity to π-stacking perturbations, in novel biosensors capable of sensitive detection of single-base mismatches, lesions [10], and a variety of DNA-protein interactions [11]. We are now asking how Nature might exploit this remarkable yet highly sensitive property of DNA, particularly since DNA-mediated CT is not limited to systems designed by chemists; charge transport also occurs in biological milieus, including the nucleosome and cell nucleus [12, 13]. Here, in addition to long-range oxidative damage and repair, DNA CT may be harnessed by proteins to accomplish cellular tasks, such as mismatch recognition [14], that we are just beginning to discover.

Despite its obvious fundamental, technological, and physiological significance, the mechanisms of DNA CT remain unclear. Understanding *how* CT proceeds through DNA is of importance, both for a complete description of the reactivity of this biomolecule and because DNA represents a paradigm for studies of CT in donor-bridge-acceptor systems; it is a highly defined, yet conformationally dynamic, molecular π-stack. Our understanding of the mechanistic features that govern DNA CT will also guide the rational design of molecular materials that exploit DNA CT and rational searches for when, where, and how Nature might harness DNA CT.

The current mechanistic paradigms, superexchange-mediated tunneling and incoherent hopping of localized charge [15], provide incomplete descriptions of DNA CT even within designed chemical assemblies, let alone in complex cellular environs. Models relying solely on the relative energies of the DNA bases fail to

Charge Transfer in DNA: From Mechanism to Application. Edited by H.-A. Wagenknecht

ISBN: 3-527-31085-1

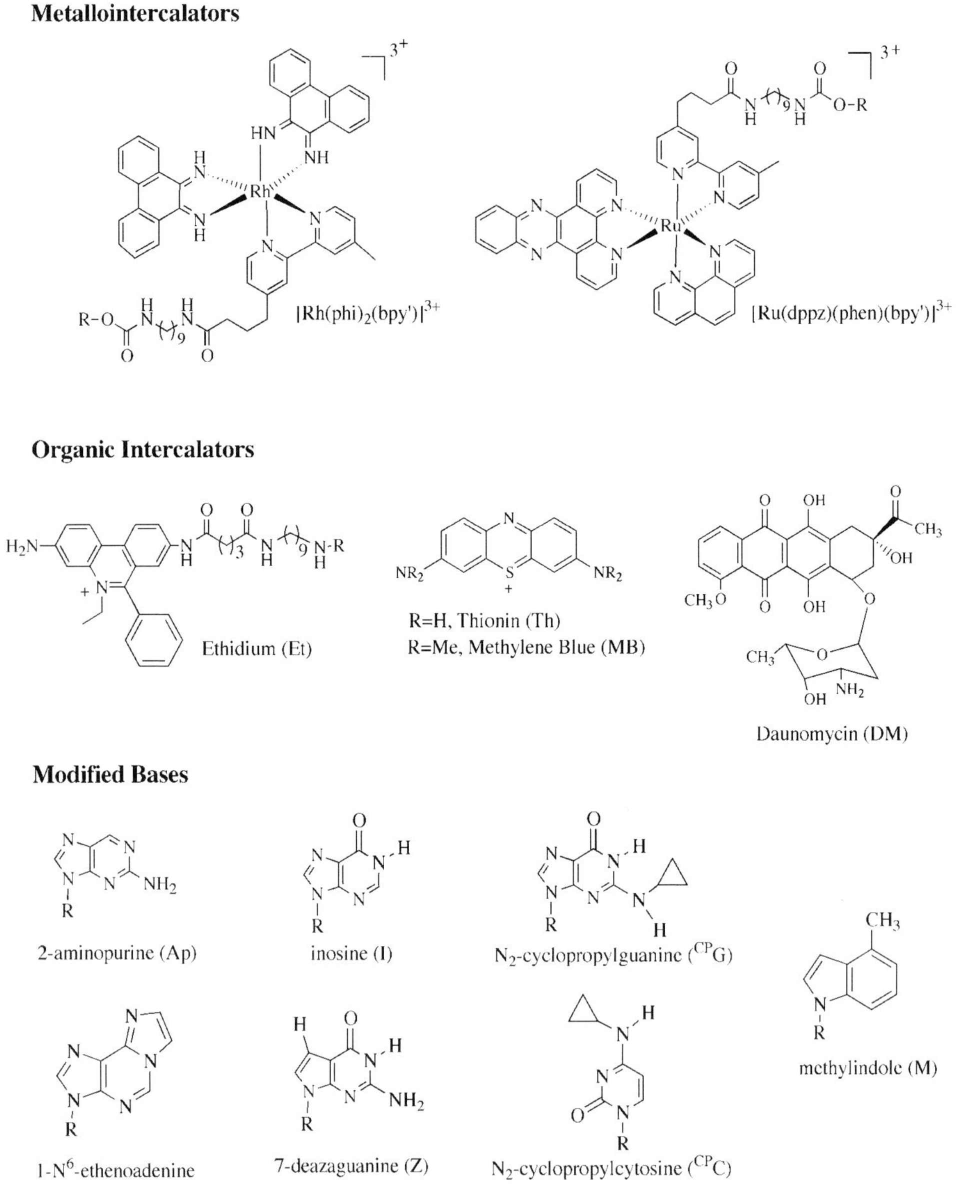

Fig. 2.2 Structures of intercalators and modified DNA bases used in studies of DNA charge transport.

2.2.1.1 Metallointercalators

Several metallointercalators are used to study long-range CT in DNA, notably phenanthrequinone diimine (phi) complexes of Rh(III) (e.g., $[Rh(phi)_2bpy']^{3+}$) and dipyridophenazine (dppz) complexes of Ru(II) (e.g., $[Ru(phen)(bpy')(dppz)]^{2+}$) [22]. These ligands facilitate tight binding ($K \geq 10^6\ M^{-1}$) of the octahedral Rh(III) and Ru(II) complexes to DNA by intercalation, and the interactions of these metallointercalators with DNA have been extensively characterized [23]. In addition, we have devised facile synthetic schemes for covalent attachment of these complexes to DNA [24]. Due to their positive charge, CT reactions between these metallointercalators and DNA bases are charge shift reactions.

Phi complexes of rhodium bind DNA avidly by intercalation of the phi ligand [25, 26]. High-resolution NMR studies indicated that binding is from the major groove with a sequence specificity that is modulated by the ancillary, non-intercalating ligands [27, 28]. This binding mode was confirmed by a 1.2-Å crystal structure of the Δ-α-$[Rh(R,R\text{-dimethyltrien})phi]^{3+}$ complex intercalated into a DNA octamer [29]. Five independent views of the octamer also established that intercalation induces no significant local or global perturbation of the DNA π-stack. No kinking or bending is associated with intercalation; the phi ligand is accommodated by a slight unwinding of the helix and can essentially be described as an additional base pair.

The photochemistry of phi complexes of rhodium makes them particularly useful probes of DNA CT [30]. When these complexes are excited with ultraviolet light ($\lambda = 313$ nm), a hydrogen atom abstraction from the sugar leads to direct scission of the DNA sugar-phosphate backbone, marking the binding site of the complex. Alternatively, exciting these complexes with visible light ($\lambda \geq 365$ nm) generates a potent photooxidant ($E^0(Rh^{3+*/2+}) \sim 2$ V vs. NHE) [31] which, when intercalated into DNA, leads to oxidative damage at guanines. The excited-state oxidation potential of these phi complexes should be sufficient to oxidize all four DNA bases as well as thymine dimers.

Dipyridophenazine complexes of Ru(II) have been dubbed "DNA light switches" [32, 33]. The luminescence of these complexes observed in organic solvents is quenched in aqueous solution as a result of proton transfer from water to the nitrogen atoms of the phenazine [34]. Intercalation within the DNA π-stack protects the phenazine nitrogen atoms from water, allowing these molecules to luminesce strongly when bound to DNA. NMR investigations of partially deuterated $[\Delta\text{-}Ru(phen)_2dppz]^{2+}$ bound to a DNA hexamer indicated that the dppz ligand intercalates from the major groove with two distinct binding modes [35]. In one orientation the phenazine nitrogens are protected from solvent, whereas in the other orientation they are more solvent exposed. Thermodynamic studies reveal tight binding to DNA [36].

Photoexcitation of Ru(II) dppz complexes generates a metal-to-ligand charge transfer excited state that is localized on the dppz ligand. This directs the charge transfer into the π-stack rather than onto an ancillary ligand [37]. We exploit the redox properties of photoexcited Ru(II) ($E^0(Ru^{2+*/3+}) \sim 0.6$ V vs. NHE) [38, 39] to generate a potent ground-state Ru(III) oxidant ($E^0(Ru^{3+/2+}) \sim 1.6$ V vs. NHE) [38, 40] in situ by the flash-quench technique [41]. This technique, originally designed for exploring electron transfer in proteins [42], has proven invaluable for the generation of powerful ground-state oxidants bound to DNA. Here, the excited state complex, Ru(II)*, is quenched by electron transfer (ET) with a non-intercalated oxidant (e.g., methyl viologen, $Ru(NH_3)_6$), yielding the ground-state Ru(III) oxidant intercalated within the DNA π-stack. The diffusion-controlled reaction required to generate the Ru(III) oxidant sets the time scale for direct measurement of reaction kinetics (typically $k \sim \leq 10^7 - 10^8\ s^{-1}$). The Ru(III) dppz complexes are sufficiently potent to oxidize guanine, and perhaps adenine, based on the potential of these nucleosides in solution.

2.2.1.2 **Organic Intercalators**

Intercalating organic molecules have also contributed to our efforts to understand DNA-mediated CT. The intercalative interactions and fluorescence properties of ethidium (Et) in the presence and absence of DNA have been extensively characterized. Ethidium intercalates tightly into DNA ($K \sim 10^6\ M^{-1}$) [43], and its luminescence intensity is significantly enhanced upon DNA binding [44]. The excited state of Et, formed upon irradiation with visible light, has a potential that is insufficient to oxidize the natural DNA bases ($E^0(Et^{*/0})$ = 1.2 V vs. NHE) [45]. This allows selective reaction with appended redox reagents or modified bases of low oxidation potential such as 7-deazaguanine (^{Z}G). Moreover, functionalization of Et with methylene tethers facilitates covalent attachment to DNA in order to precisely control the distance of CT [45, 46].

Other organic intercalators serve as probes for electrochemical investigations of ground-state CT through DNA films. Methylene blue (MB) is a three-ringed heterocycle that binds to DNA by intercalation ($K \sim 10^6\ M^{-1}$) with a slight preference for GC-rich sequences [47–49]. MB can be reversibly reduced at negative potentials ($E^0_{red} \sim -0.25$ V vs. NHE). Daunomycin (DM) is also reversibly reduced at negative potentials ($E^0_{red} \sim -0.4$ V vs. NHE). Cross-linking of DM to the exocyclic amine of guanine generates a DNA duplex with a covalently appended redox probe that has been crystallographically characterized [50, 51]. MB, DM, and Et are positively charged, and therefore charge transfers involving these intercalators are charge shift reactions.

2.2.1.3 **Modified Bases**

We use modified DNA bases to tune the redox properties, chemical reactivity, and photophysics of the bases with minimal impact to DNA structure and dynamics. Oligonucleotides containing base analogues at defined positions are readily prepared by solid-phase synthesis. The hydrogen bonding and stacking interactions of base analogues within DNA duplexes are often comparable to the natural bases and may be characterized by a variety of spectroscopic tools. Crystallographic or NMR structures of several DNA duplexes possessing base analogues now exist. The base analogues we employ are neutral molecules, and their CT reactions with natural DNA bases involve charge separation.

By using base analogues, it is possible to modulate the redox potential within the DNA π-stack with minimal structural perturbation. Examples are 7-deaza-guanine (^{z}G) and 7-deaza-adenine (^{z}A), which are derived by replacing the N-7 atom of guanine and adenine, respectively, by a C-H. This trivial structural modification lowers the oxidation potential by ~ 300–400 mV [45, 52, 53]. Alternatively, inosine (I), a guanine analogue lacking the exocyclic amine, is ~ 200 mV more difficult to oxidize [54]. While substitution of I for G in duplex DNA results in the loss of one hydrogen bond to cytosine, other base analogues form no hydrogen bonds with natural bases but instead are stabilized by base-stacking interactions [55]. An example of particular utility in our CT investigations is methylindole (M). This molecule has an exceptionally low redox potential ($E_0^{(0/+)}$ ~ 1 V vs. NHE) [56] and thus can serve as a relatively deep hole trap. In addition, since the methylindole radical cation exhibits significant absorption in the visible region [57], M is an ideal probe for investigations of CT by transient absorption spectroscopy.

Other substitutions have relatively little impact on redox potential but instead alter the reactivity of the base analogue. For instance, guanine, although a low-energy site, is relatively slow (milliseconds) to react with water and/or oxygen in duplex DNA [41, 58]. Thus, experiments monitoring permanent oxidative lesions at G may not always reflect CT chemistry occurring on a much faster time scale. Consequently, N_2-cyclopropylamine-substituted bases have been developed as kinetic traps for holes residing on G [59], A [60], and C [61]; rapid ring opening of the cyclopropylamine radical cation significantly accelerates the rate of trapping (pico- to nanoseconds) [62]. The substitution of a single cyclopropyl ring on the exocyclic amine does not significantly alter the redox properties or the hydrogen bonding, stacking, and structure of DNA duplexes containing these analogues.

The adenine analogues 1-N^6-etheno-adenine (εA) and 2-aminopurine (Ap) further illustrate the utility of modified bases as probes, particularly those with photophysical properties distinct from the natural DNA bases. The solution structures of DNA duplexes containing εA [63] and Ap [21 a] have both been characterized by NMR. Opposite thymine, Ap is base paired and stacked in duplex DNA in a manner that is very similar to adenine. Conversely, εA assumes a non-planar conformation and exhibits correspondingly poorer stacking interactions with neighboring bases. These probes can be selectively excited in DNA, generating photooxidants (Ap: $E^{0(*/-)}$ ~ 1.5 V; εA: $E^{0(*/-)}$ ~ 1.4 V vs. NHE) that oxidize distant bases with rate constants, efficiencies, and distance dependences that correlate with stacking [54]. These CT reactions are readily monitored by fluorescence quenching; although the natural DNA bases are essentially non-fluorescent [64, 65], both εA [66, 67] and Ap [68] emit strongly in solution and in DNA.

Emission from these photoexcited molecules, particularly Ap*, is remarkably sensitive to the DNA environment [21]. Consequently we have developed Ap* into a dual reporter of structural dynamics and CT within nucleic acid assemblies [69]. We have exploited the fast trapping at N_2-cyclopropylguanine to obtain direct chemical evidence for oxidation of distant guanines by Ap* [70]. To examine the initial step by fluorescence quenching, we distinguish CT with G by comparing redox-active duplexes to otherwise identical reference duplexes where the G hole donor is

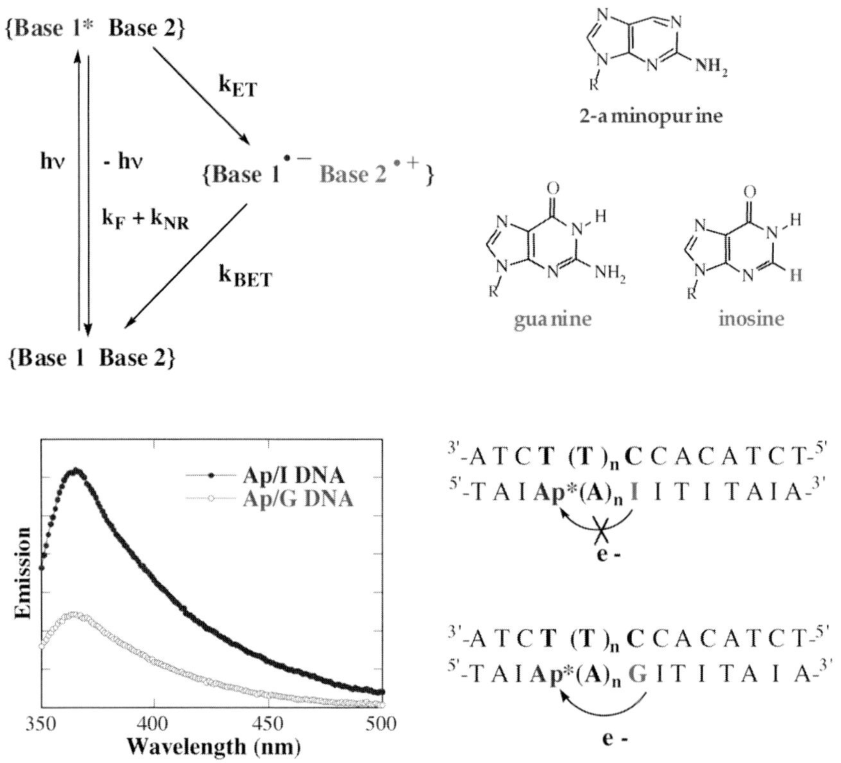

Fig. 2.3 Illustration of base-base electron transfer. Excitation of the base analogue 2-aminopurine (A_P) leads to fluorescence quenching when the easily oxidized guanine base, but not inosine, is present.

replaced by I (Figure 2.3). Due to its higher oxidation potential (E_{ox} ~ 1.5 V vs. NHE) [54], I is much less reactive towards CT with Ap* [4]. The rate constant for CT between Ap* and G is defined as $k_{CT} = k_G - k_I$, while the yield of CT (Fq) is evaluated from fluorescence quantum yields (Φ) as Fq $= 1 - \Phi_G/\Phi_I$, where Φ_G/Φ_I is the relative fraction of G-containing duplexes fluorescing and thus not undergoing CT. Importantly, for CT through intervening DNA bridges, localized injection onto the bridge (i. e., as in an incoherent hopping mechanism) does not contribute to k_{CT} or Fq, since this pathway exists in both redox-active and reference duplexes. Thus, k_{CT} and Fq describe CT via mechanisms involving superexchange and/or injection of delocalized charge.

2.2.2
Spectroscopic, Biochemical, and Electrochemical Approaches

Our spectroscopic investigations exploit photon energy to initiate CT between a photoexcited donor and an acceptor positioned elsewhere in the DNA helix.

Through a variety of spectroscopic techniques, we have observed the time-resolved dynamics of CT from femtoseconds to milliseconds. Time-resolved investigations are coupled to steady-state spectroscopy in order to evaluate both reaction rate constants and yields. Fluorescence spectroscopy is employed to measure excited-state lifetimes of redox probes within the DNA π-stack and to monitor CT quenching of these photoexcited reagents; this defines the time scale and efficiency of the reaction. With transient absorption spectroscopy, it is further possible to directly observe transient intermediates along the CT pathway and to monitor back electron transfer (BET).

Unlike spectroscopic methods, biochemical assays cannot detect intermediates, nor can they distinguish among charge injection, migration, and trapping. Instead, biochemical methods establish and probe "chemistry at a distance" by monitoring DNA damage and repair products generated by long-range DNA-mediated CT. A typical experiment involves steady-state irradiation of a photooxidant positioned site-specifically in DNA, leading to injection and migration of a hole through the duplex. These holes are ultimately trapped at the sites of lowest oxidation potential, which are guanine sites, often multiple guanines, in native DNA. Selective oxidation at the 5′-G of guanine doublets has emerged as a signature of CT-induced damage, consistent with theoretical predictions that the HOMO of guanine doublets is localized on the 5′-G [71]. Nonspecific oxidation of both guanines is indicative of alternative chemistry, such as reaction with singlet oxygen. The guanine radical cation or radical generated by DNA-mediated CT can be trapped by water and/or oxygen on a much slower time scale, yielding permanent lesions that are revealed as strand breaks upon enzymatic treatment or reaction with base (e. g., piperidine) [58]. The resultant fragments are separated, visualized, and quantified by polyacrylamide gel electrophoresis (PAGE) or high-performance liquid chromatography (HPLC). Alternatively, kinetic hole traps such as N_2-cyclopropylamine-substituted bases can be utilized to accelerate the trapping reaction and reveal CT chemistry occurring on shorter time scales [59–61].

We have also devised new methods for electrochemical probing of DNA-mediated CT [72, 73]. Through molecular self-assembly, we generate well-defined monolayers of thiol-modified DNA duplexes on gold electrodes. A redox-active intercalator, such as MB or DM, bound to the DNA at a distance from the gold surface acts as a reporter of electrochemically initiated CT through the intervening DNA bridge. The yield of reduced intercalator, monitored by, for instance, cyclic voltammetry or chronocoulometry, provides a measure of the efficiency of CT through DNA. Since the charge must traverse the alkane-thiol linker in a rate-limiting process [74, 75], before accessing the DNA π-stack, it may not be possible to obtain direct information on the rates of CT through these DNA films. Note also that in contrast to other techniques, the electrochemical methods probe electron, rather than hole, transport between ground states, rather than photoexcited molecules.

2.3
Understanding the Fundamental Parameters Governing DNA-mediated Charge Transport

In the following sections, we discuss the parameters that have emerged from our work as being most significant to the rate constants, yields, and distance dependences of DNA CT.

2.3.1
The Base Pair π-Stack of Double-helical DNA Regulates Charge Transport

2.3.1.1 Spectroscopic Investigations of Charge Transport through DNA

That DNA-mediated CT requires strong coupling of the redox probes to the π-stacked base pairs was evident even in our earliest investigations. Our initial studies found enhanced efficiency of photoinduced electron transfer between polypyridyl metal complexes when weakly bound to DNA; the rate constant for luminescence quenching of $Ru(phen)_3^{2+}$ (phen = 9,10-phenanthroline) by tris(phenanthroline) complexes of Co(III) and Rh(III) increased by two orders of magnitude [76, 77]. The importance of coupling to the DNA π-stack was revealed by our metallointercalator redox probes, which bind DNA avidly by intercalation between the bases [31, 39, 78]. Oxidative quenching of dipyridophenazine (dppz) complexes of Ru(II) by phenanthrequinone diimine (phi) complexes of Rh(III) was observed to be extremely rapid ($k > 10^{10}\ s^{-1}$) only when both complexes were intercalated into the DNA base stack. Monitoring DNA-mediated quenching of dppz complexes of Ru(II) or Os(II) by $Rh(phi)_2bpy^{3+}$ (bpy = 2,2′-bipyridine) by transient absorption afforded direct observation of the oxidized Ru(III) or Os(III) intermediate, thus confirming CT as the quenching mechanism [38, 79]. Moreover, in the first investigation of DNA CT over a precisely defined distance between covalently appended redox probes [2], we reported that luminescence of a photoexcited Ru(II) intercalator was quenched by a Rh(III) intercalator over 40 Å away. Significantly, non-intercalating, tethered Ru(II) and Rh(III) complexes did not undergo this quenching reaction. The importance of strong coupling to the DNA π-stack, as achieved through intercalative stacking, to rapid and efficient CT was thus definitively demonstrated.

That strong coupling of the redox probe within the DNA base stack is requisite for DNA CT suggests that the π-stacked base pairs provide the pathway for charge migration. Consequently, one would also expect coupling between the bases in duplex DNA to be critical for DNA-mediated CT. A dramatic demonstration that this is indeed the case comes from our observations of the influences of single-base mismatches, arguably the most subtle π-stack perturbations, on DNA CT. Our spectroscopic investigations of DNA CT have consistently revealed that a single-base mismatch markedly diminishes the yield of CT. This was first seen for CT between photoexcited Et and $[Rh(phi)_2bpy']^{3+}$ covalently appended to DNA and intercalated within the base stack [46]. As with the metallointercalators, rapid (10 ns) CT quenching of Et* luminescence by the Rh(III) intercalator over distances of

20–30 Å was observed in fully matched DNA; the yield of CT exhibited a very shallow dependence on distance, while the fluorescence decay rate of Et* was unaffected by the position of the Rh(III) quencher. Remarkably, however, the incorporation of a single C-A mismatch in the intervening DNA bridge significantly reduced the yield of CT. This result confirmed that the pathway of CT is through the base stack, not via the sugar-phosphate backbone, and revealed the exquisite sensitivity of ultrafast DNA-mediated CT to base stack structure and dynamics.

Similar results were obtained for DNA-mediated CT quenching of Et* luminescence by the base analogue 7-deazaguanine [45]. In this case, a DNA base acts as a direct participant in long-range CT. Using steady-state fluorescence and nanosecond time-correlated single-photon counting (TCSPC), efficient CT quenching was observed to proceed on a sub-nanosecond time scale over long molecular distances (up to 27 Å). While the yield of CT again exhibited a shallow distance dependence, both the yield and the distance dependence were dramatically influenced by subtle changes in base sequence, including intervening mismatches. These data were consistent with the notion that the nature of the intervening DNA π-stack, rather than the donor-acceptor separation, governs the yield of DNA CT.

The influence of a single-base mismatch on the yield of DNA CT is detected not only by monitoring luminescence quenching but also directly through the formation of CT intermediates by transient absorption spectroscopy. Here we exploit the flash-quench scheme to generate the potent ground-state Ru(III) oxidant in situ; the intercalated Ru(III) can then oxidize guanines or other bases in DNA in reactions that we can monitor by transient absorption [41]. For instance, the neutral guanine radical is detected immediately following nanosecond laser photolysis and flash-quench generation of Δ-Ru(phen)$_2$(dppz)$^{3+}$ intercalated within poly(dG-dC). In fact, formation of the guanine radical was found to be concomitant with quenching of photoexcited Δ-Ru(phen)$_2$(dppz)$^{2+}$ by the diffusible quencher ($k_{obs} \sim 2 \times 10^7$ s^{-1}); oxidation and deprotonation of guanine occur in less than ~200 ns in duplex DNA.

In analogous experiments we have directly observed a radical cation intermediate localized on a DNA base analogue (Figure 2.4) [56, 80]. Here the flash-quench generated and intercalated Ru(III) oxidizes methylindole, and the resultant M radical cation can be detected by both transient absorption and EPR spectroscopies. These data provided the first direct measurements of rates of formation of transient radical species over long (>20Å) molecular distances and can be correlated with biochemical assays of permanent oxidative damage at the M site. As for guanine, formation of the CT intermediate was concomitant with quenching of the photoexcited Ru(II) intercalator. In fact, for the assemblies examined, containing no intervening guanines, CT to form the methylindole radical cation was not rate limiting over a distance of at least ~40 Å. However, the incorporation of a single C-A mismatch essentially eliminated the formation of the M radical cation and subsequent damage products [80]; as observed in experiments monitoring luminescence quenching, and here when detecting the formation of CT intermediates, single-base mismatches can effectively "turn off" CT.

It has also been possible to detect the influence of variations in base stacking by examining CT between DNA bases. In this effort we utilize non-natural DNA bases, and/or to the different time scales and chemistries probed in measurements of rate constants for HT versus yields of permanent reaction. To address directly the role of directional asymmetry, we investigated base-base HT from 5′-3′ and 3′-5′ in duplex DNA.

Using steady-state fluorescence, we first examined the yield and distance dependence of HT between Ap* and G through $(A)_n$ bridges, where n = 0–4 [85]. Remarkably, for HT reactions over the same distance, through the same intervening DNA bridge, and with the same net driving force, we observed that HT is more efficient and less distance dependent from 5′-3′ than from 3′-5′. Yet, while these experiments are a dramatic demonstration of the sensitivity of DNA CT to base stack structure, they do not reveal *how* directional asymmetry influences HT. Does directional asymmetry modulate only the relative population undergoing HT (i. e., the yield), as observed for single-base mismatches, or does it also influence the ultrafast dynamics of HT? To answer this question we measured with femtosecond resolution the ultrafast dynamics of base-base HT from 5′-3′ and 3′-5′ in duplex DNA [86]. Significantly, the rate constant for HT between Ap* and G is also dependent on direction; HT between Ap* and G when in contact, and when separated by an intervening A bridge, is faster from 5′-3′ than from 3′-5′. Although we have observed many times the exquisite sensitivity of the yield of DNA-mediated CT to perturbations in the base stack, such as mismatches, these data indicate that subtle variations in the base stack also modulate the dynamics of DNA-mediated CT. The influence of directional asymmetry on both the yields and rate constants for HT can be accounted for by differences in electronic coupling and driving force, as both are highly sensitive to base stack structure and dynamics. Additionally, directional asymmetry in base step dynamics may influence the rate of rearrangements necessary for conformationally gated CT (*vide infra*).

2.3.1.2 **Biochemical Investigations of Long-range Oxidative Damage**

Our discussion of the role of base stack structure and dynamics has focused thus far on the initial steps of DNA CT monitored primarily by fluorescence and transient absorption spectroscopy. However, the fact that DNA can act as a conduit for rapid and long-range CT has lead to questions regarding the consequences of charge migration and the fate of electrons and holes in DNA. What is the physiological relevance of these rapid, long-range CT reactions? How does DNA-mediated CT induce chemistry, particularly oxidative damage, from a distance? As with our spectroscopic studies, our biochemical assays of permanent oxidative damage via DNA-mediated CT reveal chemistry that is relatively insensitive to distance but strongly modulated by the integrity of the intervening DNA bridge.

Our biochemical experiments probe the long-range migration of electron holes through DNA. As guanine has the lowest oxidation potential of the isolated natural bases in solution [87], a hole migrating through duplex DNA is expected to be ultimately trapped at guanine sites, leading, potentially, to permanent oxidative damage. Indeed, our first investigations confirmed that DNA-mediated CT induces chemistry at a distance, namely, long-range oxidation of guanine [8, 88].

However, the redox potentials of the DNA bases are likely to be influenced by stacking interactions with neighboring bases in the helix, and oxidative damage patterns may reflect these sequence-dependent variations. This is in fact the case; multiple guanines such as doublets or triplets are particularly susceptible to permanent oxidation. The 5′-G of 5′-GG-3′ doublets is especially sensitive [89], and its tendency to be preferentially oxidized has become a characteristic signature of CT chemistry. The most prevalent rationale for this oxidative damage pattern is the asymmetrical distribution of the HOMO in DNA as revealed by calculations on isolated gas-phase base steps [71].

We first examined the chemistry of migrating holes in DNA using DNA assemblies containing a tethered intercalating photooxidant, $[Rh(phi)_2DMB]^{3+}$, spatially separated from two GG sites [8]. Photolysis at 313 nm confirmed that the rhodium complex intercalated near the terminus of the duplex where it was tethered and that no inter-assembly intercalation had occurred. Irradiation with 365-nm light induced long-range oxidation of both GG sites 17 Å and 34 Å away from the photooxidant; remarkably, the yield of oxidative damage was essentially the same at both GG sites. The damage yields were, however, very sensitive to changes in oxidation potential at the guanine sites or to changes in stacking of the intercalator and of the intervening DNA bases. Significantly, DNA assemblies possessing base bulges between the 5′-GG-3′ sites distal and proximal to the intercalated photooxidant exhibited a dramatic reduction in the distal/proximal ratio of oxidative damage [88]. These seminal experiments thus established that DNA-mediated CT leads to long-range oxidative damage via hole migration *through* the DNA base stack.

Consistent with our spectroscopic studies of DNA CT, this long-range oxidative chemistry exhibits a pronounced sensitivity to stacking with minimal dependence on distance. In fact, our systematic studies of the distance dependence of long-range oxidative damage by a tethered Rh(III) intercalator revealed no significant attenuation in the yield of oxidation over a distance of 75 Å [6]. Moreover, long-range guanine oxidation occurs at distances at least 200 Å away from the site of hole injection [6, 90] (Figure 2.6). Even over this biologically relevant range, distance was not the primary factor regulating the yield of oxidation. Instead the intervening sequence and sequence dynamics were found to be critical [6]. For instance, the yield of oxidative damage was diminished through multiple 5′-TA-3′ steps, and an increase in the proportion of long-range damage was observed as the temperature was increased from 5 °C to 35 °C.

Oxidation of guanine via long-range DNA-mediated CT is not specific to our Rh(III) intercalators. We have observed this chemistry with several other intercalating oxidants, including ground-state Ru(III) species [6, 91] and ethidium [92]. Indeed, the generality of this DNA-mediated chemistry at a distance has now been established by investigations of several different laboratories using a variety of distinct oxidants [1b–d, 59, 83]. Yet, while DNA clearly possesses an inherent ability to mediate long-range CT, since the base pair stack is the pathway for transport, the rate constants, yields, and distance dependence of CT are tightly regulated by base stack structure and dynamics.

We prepared DNA duplex-quadruplex conjugates to probe long-range guanine oxidation in guanine quadruplexes [108]. A Rh(III) photooxidant covalently bound to the duplex end was found to induce oxidative damage to distant guanines within the quadruplex structure. Two features of oxidative damage within the quadruplex were particularly striking. First, the damage patterns at repetitive guanines within the quadruplex were distinct from that observed in B-DNA duplexes. Specifically, G damage in the quadruplex occurred almost exclusively at the external tetrads; limited oxidation of the internal G sites likely reflects protection of these sites from trapping. Second, guanine oxidation in the quadruplex was more efficient than oxidation of G doublets in the B-DNA duplex. These observations are particularly significant to analyses of oxidative damage within cells, especially given that quadraplexes may form at the end of telomeres [109] and given the proposal that oxidative damage is funneled to telomeric regions within the genome [110].

While DNA polymorphism affords a variety of non B-form base stacks, the interaction of B-DNA with physiological partners, such as proteins, may also strongly affect base stack structure and dynamics. Indeed, several investigations have revealed that DNA-binding proteins can significantly modulate the efficiency of long-range guanine oxidation [111, 112]. In these experiments we employ DNA assemblies containing a protein-binding site between two guanine doublets spatially separated from a tethered $[Rh(phi)_2bpy']^{3+}$ photooxidant. DNA-binding proteins such as M.*Hha*I or TBP, which disrupt DNA base stacking by base flipping or dramatically kinking the DNA, respectively, decrease the yield of guanine oxidation upon binding. Remarkably, however, long-range CT chemistry is recovered with the mutant M.*Hha*I (Q237W) protein that inserts an aromatic tryptophan residue within the DNA, thereby restoring the π-stack. Alternatively, proteins such as the restriction endonuclease *R.Pvu*II and the transcription factor Antennapedia homeodomain protein (ANTP), which do not disrupt the π-stack but may rigidify the helix upon binding, enhance the ability of DNA to mediate long-range CT.

These results, completely analogous to our electrochemical assays (*vide infra*), illustrate how protein binding sensitively modulates long-range DNA-mediated CT in a manner that is very specific to the nature of the protein-DNA interactions. Consequently, we have applied this chemistry as a probe of protein-DNA interactions, such as that between MutY, a DNA repair enzyme found in *Escherichia coli*, and DNA [113]. In this case, no disruption of DNA-mediated CT is detected upon protein binding, suggesting that MutY does not locate its binding site by progressive base flipping.

Our biochemical assays of long-range guanine oxidation, and our spectroscopic investigations of earlier steps in the CT process, have made it abundantly clear that a defining feature of DNA-mediated CT is its precise regulation through base stack structure and dynamics. Perhaps this should not be too surprising, given that all known functions of DNA exploit its rich, sequence-dependent flexibility. Yet, current paradigms of DNA CT cannot predict, or rationalize, this defining feature of DNA CT. Certainly such a definitive parameter must be incorporated as a central feature in realistic models of DNA CT.

2.3.2
The Role of the Oxidant in DNA-mediated Charge Transport: Energetics, Coupling, Lifetimes, and Back Electron Transfer

Rapid and long-range DNA-mediated CT has now been observed in a variety of assemblies initiated by a diverse array of ground- and excited-state oxidants including metallointercalators e.g., [2, 8, 41], organic intercalators e.g., [3, 114], end-capped molecules e.g., [1c, 115], derivatized nucleotides e.g., [52, 82a], base analogues e.g., [54], and even sugar radicals e.g., [116]. Undeniably it is an intrinsic feature of DNA itself. However, it is also apparent that not all oxidants are created equal; variations in the rate constants, yields, and distance dependences are evident among all of these oxidants. To examine this issue, it is instructive to consider DNA CT in terms of an initial charge injection, followed by subsequent CT events. In this way we can ask how the oxidant influences (1) injection of charge into the DNA assembly and (2) charge migration and subsequently long-range oxidative damage.

2.3.2.1 Rate Constants and Net Yields of Charge Injection

We may consider how characteristics of the oxidant influence charge injection into DNA, or a donor-bridge-acceptor (D-B-A) assembly, in general. Important is the injection free energy, ΔG_{inj}, i. e., the minimum free energy separation between a state with a hole localized on the DNA bridge and the initial state, i. e., D*-B-A. For DNA systems, this gap is relatively small compared to, for instance, proteins. Consequently, DNA CT, particularly its distance dependence, is more sensitive to the properties of the oxidant. Variations in ΔG_{inj} with different oxidants exert two effects. First, for oxidants that are lower in energy than the oxidized bridge states, superexchange-mediated tunneling is typically invoked to rationalize transport to a distant hole acceptor. In such a mechanism, the migrating hole does not actually reside on the DNA bridge, and the rate constant for CT decays exponentially with distance (r), governed by β, a parameter that reflects the falloff in electronic coupling with distance for a given intervening bridge:

$$k_{superexchange} \propto \exp(-\beta r) \tag{1}$$

Here, the injection barrier influences the rate constants and distance dependence of CT by tuning β; β becomes smaller as ΔG_{inj} decreases [115]. Second, in the limit of very small or negative values for ΔG_{inj}, for oxidants that are close to or above the oxidized bridge states, a transition from a superexchange to a hopping-type mechanism is expected. Here, charge actually occupies the DNA bridge, and the rate constant for CT no longer decays exponentially with distance. Instead, it falls off algebraically, with the number of hopping steps (N) modulated by the parameter η ($1 \leq \eta \geq 2$):

$$k_{hop} \propto N^{-\eta} \tag{2}$$

Consequently, the energetics of the oxidant tunes β for superexchange and may determine whether charge migration involves bridge occupancy, as in a hopping mechanism, or tunneling. Certainly these oxidant-dependent variations may lead to significant distinctions in the rate constants, yields, and distance dependence of CT through DNA.

Yet, energetics of the oxidant is certainly not the only factor, or even the decisive factor, for charge injection into DNA. As an example, consider two types of DNA assemblies where ultrafast rate constants for CT have been obtained using femtosecond fluorescence and transient absorption spectroscopy. In DNA constructs containing a covalently tethered, intercalated Et, photoexcitation of Et induces oxidation of ^{Z}G through mixed-sequence DNA bridges; this long-range CT reaction was characterized by a distance-independent rate constant, k_{CT}, of 2×10^{11} s^{-1} for donor-acceptor separations between 10 Å and 17 Å [3]. Conversely, k_{CT} for base-base CT between Ap* and G through intervening $(A)_n$ bridges, although equally rapid over short distances ($k_{CT} = 1 \times 10^{11}$ s^{-1} for n = 0), was found to exhibit a weak exponential decay with increasing donor-acceptor separation (β ~ 0.6 $Å^{-1}$ if a superexchange mechanism is operative) [4]. Yet the free energy for injection of charge by Et* and Ap* onto their respective DNA bridges is similar (± ~100 mV); if anything, Ap* is slightly above its DNA bridge, while Et* is slightly below. Thus, arguments based solely on oxidant energetics would predict a similar, or even shallower, distance dependence for CT initiated by Ap*.

More important here may be the interactions of the photooxidants with the π-stacked base pairs in duplex DNA. As a base analogue in B-DNA, Ap* is largely limited to interactions within its own strand and to conformational states typical of DNA bases. In contrast, Et*, with its large heterocyclic aromatic surface area for intercalation, is intimately coupled with DNA bases on both strands. At the same time, however, Et* also exhibits significant and rapid molecular motions when intercalated in duplex DNA (τ ~ 75 ps). The combination of strong coupling with flexibility may be crucial for achieving conformations optimally active for CT reactions. Significant in this context is the fact that we observe CT between Ap* and G over longer distances at elevated temperatures where access to "CT-active" conformations may become more facile (*vide infra*). Similarly, the exceptional rigidity of DNA hairpin assemblies constrained by a terminal stilbene oxidant may contribute to the steep distance dependence of DNA CT observed in these constructs [117].

Consistent with this important role of oxidant-DNA coupling, we have repeatedly observed that the ability of an oxidant to initiate damage at a distance is ultimately related to its association with the DNA base stack. This was unambiguously demonstrated in a study of oxidative damage by a family of Ru(II) complexes [40]. While intercalating Ru(II) complexes such as $Ru(bpy)_2(dppz)^{2+}$ promote significant long-range oxidation of guanine, non-intercalating, groove-binding species such as $Ru(bpy)_3^{2+}$ are ineffective. Moreover, for a series of Ru(II) intercalators with comparable redox properties, the yield of guanine oxidation was found to be directly correlated to the strength of intercalative binding. Oxidant-DNA coupling requisite to CT can be achieved by intercalation, but it is sensitively modulated by the strength and nature of the interactions between the oxidant and the DNA bases.

Even oxidants with redox energetics and coupling similar to that of the DNA π-stack may display different net yields of charge injection. Important here is the lifetime of the initially generated radical ion–ion pair. While oxidants that are well coupled to the DNA π-stack may inject charge onto the DNA bridge very rapidly and with a high efficiency, fast back electron transfer (BET) within the initially generated radical ion/ion pair can severely diminish the amount of charge that remains (Figure 2.7). We have observed this when using both a base analogue (Ap) and a DNA intercalator (thionin [Th]) as photooxidants. In each case, charge injection is very rapid, occurring on the picosecond [4] and even femtosecond [114] time scale. Yet neither oxidant generates detectable permanent oxidative damage at guanines in duplex DNA [70, 118]. Clearly, the yields of oxidative damage need not reflect the rate constants for CT through DNA. We rationalized that BET is significantly faster than trapping of the guanine radical cation (or radical) by water and/or oxygen. This is reasonable given that these guanine-trapping reactions are thought to be very slow (~ milliseconds) [41] in duplex DNA. We further confirmed this hypothesis using N_2-cyclopropylguanine, ^{CP}G, to accelerate the trapping of the G radical cation. Here, both Ap* [70] and Th* [119] afford efficient long-range oxidative damage.

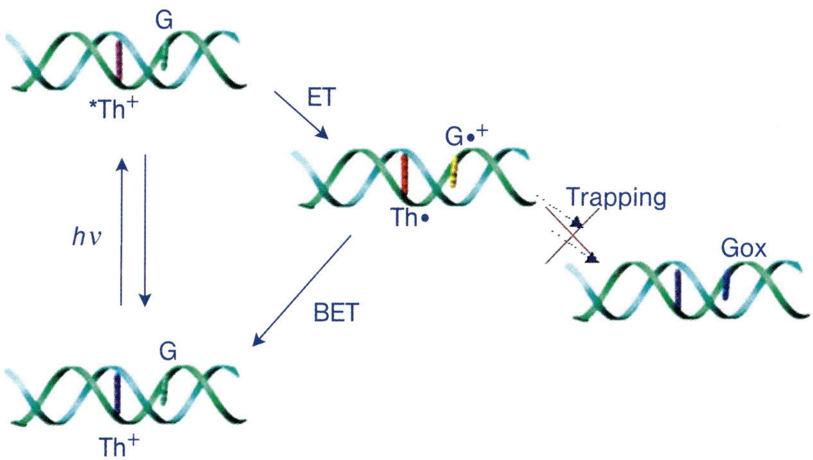

Fig. 2.7 Reversible electron transfer in thionin-DNA systems. Photoexcitation of thionin (Th) in DNA does not produce permanent guanine damage because back electron transfer (BET) is much faster than the trapping reaction for the guanine cation radical.

These results indicate that for photoinduced electron transfer (PET) in DNA, as for all PET reactions, charge recombination within the high-energy radical ion/ion pair limits product yield; the yield of net chemistry increases not only with enhanced efficiency of charge separation but also with the lifetime of the charge-separated state. For a given oxidant, the yield of permanent oxidation reflects the competition between BET and trapping and, consequently, need not always decrease with donor-acceptor separation. For instance, if the distance dependence of BET is steeper than the distance dependence of charge separation, increasing the distance between the donor and acceptor is expected to *increase* the yield of trapped product. Indeed, we observe this behavior for oxidation of ^{CP}G by both Ap* [70] and Th* [119] in duplex DNA. A similar conclusion was reached by Kawai et al. in studies of permanent oxidation at guanine in duplex DNA initiated by the intercalating photooxidant naphthalene diimide (NDI) [120]. Increasing the distance between NDI and G increased the lifetime of the NDI radical anion and concomitantly enhanced the yield of permanent guanine oxidation. Each of these studies also highlights the potential pitfalls of drawing mechanistic conclusions about DNA CT based solely on yields of permanent guanine oxidation; the relatively slow trapping at G may mask details of DNA CT occurring on a much faster time scale.

For PET between molecules in solution, the energy of the radical ion/ion pair often plays a pivotal role in the rate of BET. Aside from energetics, what controls the rate constants for BET within a radical ion–ion pair generated in duplex DNA? Certainly the base sequence surrounding the injection site must be significant. In fact, in studies where we directly monitored both the yield of transient radical ions and permanent oxidative damage, we observed a pronounced sensitivity to injection site sequence [80]. These experiments exploit a covalently tethered, intercalated Ru(III) complex, generated by flash quench, to oxidize a distant methylindole, M, within the base stack. The resulting M radical cation is readily monitored by its characteristic transient absorption, while permanent oxidation at M sites is revealed by polyacrylamide gel electrophoresis following steady-state irradiation and treatment with piperidine. We cannot resolve the rate constant for CT, even when M is ~40 Å away from the site of Ru(III) intercalation, as it is still faster than the diffusion-limited quenching reaction that generates the Ru(III) oxidant. However, the yield of the M radical cation, and permanent oxidation, was seen to depend strongly on the sequence at the injection site. Specifically, replacement of G with I at the injection site actually *increased* the yield of transient radical cation and permanent damage products. Since I is more difficult to oxidize than G, the observed influence of injection sequence is not predicated by either superexchange or thermally induced hopping models. We ascribe the enhanced yield seen with I to a reduction in the extent of BET between the reduced oxidant and the oxidized DNA [80]. With G at the injection site, the hole is presumably more localized and has a greater propensity to undergo BET. In DNA, the rate constants and yields of BET are not simply dependent on isolated donor and acceptor radical ions but are intimately related to the influence of neighboring bases on the energies and spatial distribution of these ions.

2.3.2.2 **Long-range Oxidative Damage**

While the influence of the oxidant on the initial steps of DNA CT, particularly on the injection of charge into the duplex, is perhaps intuitive, the notion that that oxidant might influence the subsequent charge migration steps is less obvious.

Yet the distance dependence and patterns of long-range oxidative damage in B-DNA are in fact distinct for different oxidants. Does this mean that the mechanisms for CT through double-helical DNA are different in these various constructs? Or, is it possible that a hole migrating through DNA has a "memory" of the oxidant that generated it? Our recent systematic investigation confirmed that the oxidant does affect long-range permanent oxidation; the migrating hole does remember its origin, and the memory mechanism is BET.

We directly compared the long-range oxidative damage induced by several oxidants commonly employed to probe DNA CT, including two metallointercalators – $Rh(phi)_2(bpy')^{3+}$ and $Ru(phen)(bpy)(dppz)^{2+}$ – and three organic intercalators – ethidium (Et), thionin (Th), and anthraquinone (AQ) [119]. Each of these oxidants was covalently tethered to one end of a DNA duplex possessing a proximal and a distal 5′-GG-3′ site separated by an intervening $(A)_6$ bridge. All assemblies were characterized by non-denaturing gel electrophoresis to confirm that none of the covalently tethered oxidants promote DNA aggregation. Our assay evaluated the quantum yield for overall damage, as well as the ratio of damage at the distal versus proximal guanine doublets. We further examined the yield of oxidation at ^{CP}G, which presents a significantly faster trap through ring opening of the ^{CP}G radical cation. These assays revealed distinct differences in long-range oxidative damage with each of the oxidants employed. Specifically, the overall damage yields varied in the order Ru > AQ > Rh > Et ≫ Th; in fact, oxidative damage by Th* could be detected only when probed with the fast ^{CP}G trap. Moreover, oxidants that afford high yields of overall damage exhibited smaller distal/proximal ratios. Thus, the distal/proximal ratios were notably lower for Ru and AQ than for Rh and Et.

The differences observed among the oxidants can be understood in terms of different rates of BET between the reduced oxidant to the oxidized DNA [119] (Figure 2.8). First consider the migration of charge between guanine doublets in duplex DNA without accounting for BET. In the limiting case where CT is rapid relative to trapping, charge can migrate throughout the duplex, or equilibrate, and ultimately be trapped at the lowest potential sites. If the potential at the two guanine

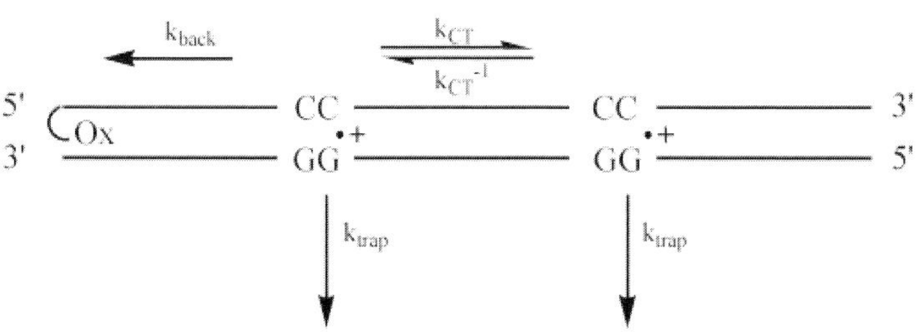

Fig. 2.8 Scheme depicting the fate of the guanine cation radical in oxidant (Ox)-DNA assemblies. Regardless of the intrinsic kinetic properties of the DNA with respect to charge transport and trapping, the rate of back electron transfer with the initial oxidant will influence the ratio of distal to proximal guanine damage.

Our investigations of ultrafast CT between Et* and ^{Z}G thus indicated that DNA-mediated CT may be gated by molecular motions of the photoexcited hole donor. This is not unexpected. Given the requirement for strong coupling of the oxidant to the DNA π-stack, it makes sense that time-dependent variations in this coupling will play a major role. Yet gating by rearrangement of the intercalator alone does not account for the distinct distance dependences observed in the rate constants versus the yields. It is also necessary to invoke gating due to conformational motions of bases in the intervening DNA bridge. For instance, assume that DNA CT requires certain favorable conformations of the bridging bases, as well as the intercalating oxidant. Increasing the number of bridging bases increases the number of molecules that must be appropriately aligned for CT, thereby decreasing the probability, and the yield, of CT.

But do conformational dynamics of the DNA bases play such a pivotal role in DNA CT? Our biochemical investigations of long-range oxidative damage found a striking correlation between the distal/proximal damage ratios and the dynamics of base pair opening at intervening mismatches [102] (*vida supra*). However these "breathing" motions occur on a much slower time scale (milliseconds) than CT through DNA. Could these dynamics also be correlated with faster base motions more relevant to charge migration? Other studies of long-range guanine oxidation proposed that structures favorable to CT may form transiently, at least on time scales sufficiently long to influence ultimate trapping at guanine [93] (*vide supra*).

To develop a more direct approach for correlating base dynamics with the rate constants, yields, and distance dependence of DNA-mediated CT, we turned to our base analogue Ap*. Given the exquisite sensitivity of Ap* photophysics to base stack structure and dynamics, and that Ap* undergoes CT reactions with other bases in DNA, it naturally emerged as our dual probe. Also advantageous is the fact that in DNA Ap behaves as a natural DNA base; it can thereby provide a realistic representation of base dynamics (on time scales within the excited-state lifetime, 10 ns) and electronic interactions between bases that are ultimately responsible for DNA-mediated CT.

We initially exploited nucleic acid polymorphism to modulate the conformational motions of the bases. Using specific Ap-containing DNA duplexes and analogous DNA-RNA hybrids, we characteristically tuned the dynamics of the base pair stack within the CT bridge [69]. Our spectroscopic investigations correlated the distance dependence of the yield of CT between Ap* and G with the conformational flexibility of the intervening bridge; a shallower distance dependence was evident in duplexes with heightened base dynamics. These results are unexpected in light of notions that dynamic disorder hinders the ability of DNA to transport charge [125]. However they are completely consistent with the proposal that CT proceeds via specific, well-coupled conformations of the DNA bases, conformations we term simply CT-active. Within a certain regime, dynamic motion of the DNA base pairs samples conformations and provides access to those that are CT-active. Thus, the number of discrete CT events may be larger for duplexes with increased conformational flexibility. In such cases, the distance dependence of the yield of CT will be shallower when compared to duplexes that have more-restricted access to these conformations.

Investigations of the temperature dependence of base-base CT probe directly how conformational motions of the DNA bases modulate DNA-mediated CT. Through femtosecond spectroscopy we examined the temperature-dependent rate constants for CT between Ap* and G in DNA assemblies where the donor and acceptor were in contact (ApG) or separated by a single intervening A (ApAG) [126]. To explicitly probe the influence of temperature-dependent base dynamics on CT in duplex DNA, 35-mer DNA assemblies with melting temperatures near 60 °C were used. The rate constants for CT were found to exhibit a pronounced temperature dependence, characterized within three temperature regimes. At lower temperatures, where the assemblies are fully duplexed, k_{CT} increased weakly with temperature, more markedly for ApG than for ApAG. Through the duplex melting regime, k_{CT} decreased dramatically, while at higher temperatures, significantly less variation in k_{CT} was evident. Consistent with these later observations were analogous experiments with single-stranded ApG and ApAG assemblies that revealed much slower decay kinetics with little temperature dependence; CT is a duplex reaction. Yet, the increase in k_{CT} with temperature in duplex DNA was much shallower than predicted by standard Marcus theory for ET. Moreover, the dichotomy between experiment and theory was greater in the ApAG duplex where CT is bridge mediated. This dichotomy is not so surprising, however, since conventional theory neglects the temperature-dependent motion of the DNA bases.

These results thus directly identified a defining role for base dynamics in DNA CT that must be included within mechanistic models. We proposed that CT does indeed require CT-active conformations: specific, well-coupled arrangements of the DNA bases, accessed through conformational rearrangements within the base pair stack. We further suggested that formation of CT-active conformations, particularly for bridge-mediated CT, gates CT and thereby governs the observed rate constants.

According to our model, the yield of CT reflects the probability of accessing CT-active conformations (Figure 2.9). We predicted that this would be strongly modulated by temperature and by the distance between the donor and acceptor, since conformational gating should become more important as the number of bridging bases increases. Consequently, we examined the temperature-dependent yield of base-base CT between Ap* and G through DNA bridges of varied length and sequence [127]. DNA assemblies (35-mers) were constructed containing adenine bridges, $Ap(A)_nG$ (n = 0–9 Å, 3.4–34 Å), and mixed bridges, ApAAIAG and ApATATG. CT was monitored through fluorescence quenching of Ap* by G, and through HPLC analysis of photolyzed DNA assemblies containing Ap and ^{CP}G, to reveal the influence of temperature on both CT and permanent oxidation. Using Ap* as a reporter of base stack structure and dynamics, we established that temperature-induced base dynamics occur on the time scale of CT (< 10 ns) and that these dynamics regulate the yield of CT as predicted by a model via CT-active conformations. For each duplex examined, the yield of bridge-mediated CT and permanent oxidation *increases* with increasing temperature-induced base dynamics.

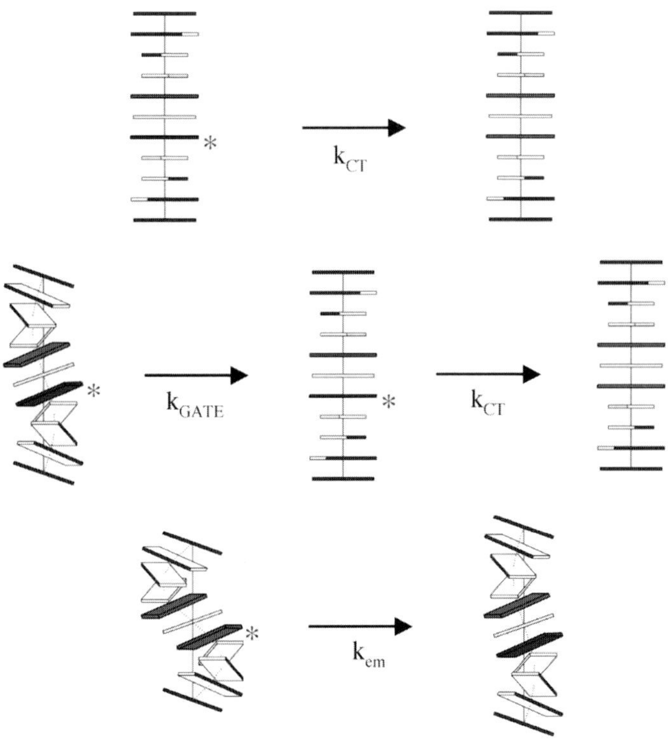

Fig. 2.9 Illustration of how fluctuations in the π-stacking of DNA affects DNA charge transport. CT is rapid in a perfectly stacked DNA (top), is gated by conformational alignment in an imperfectly stacked DNA (middle), and does not occur at all in a poorly stacked DNA (bottom), where a CT-active conformation cannot be accessed within the lifetime of the excited probe molecule.

In fact, enhanced base dynamics can extend CT to significantly longer distances, at least 34 Å. However, the influence of temperature is sensitively regulated by bridge length, becoming more dramatic for longer bridges, and by bridge sequence, paralleling variations in bridge flexibility, but not in energetics. *Though contrary to the predictions of models based on superexchange or thermally induced hopping, these results match precisely the expectation of conformationally gated CT.*

Further support for the significant role of conformational gating comes from investigations of base-base CT between Ap* and G conducted in rigid LiCl glasses at 77 K, where conformational rearrangement is effectively eliminated [128] (Figure 2.10). The yield of CT between Ap* and G through $(A)_n$ bridges (n = 0–4) was monitored by fluorescence quenching. Again using Ap* as a sensitive fluorescent probe of base stack structure and dynamics, we established that neither the high concentration of LiCl (10 M) nor cooling to 77 K significantly affects the duplex structure, including the stacking interactions and solvent accessibility of the DNA bases; the LiCl glasses provide a non-perturbing environment where conformational modes are kinetically frozen. Remarkably, DNA-mediated CT between Ap*

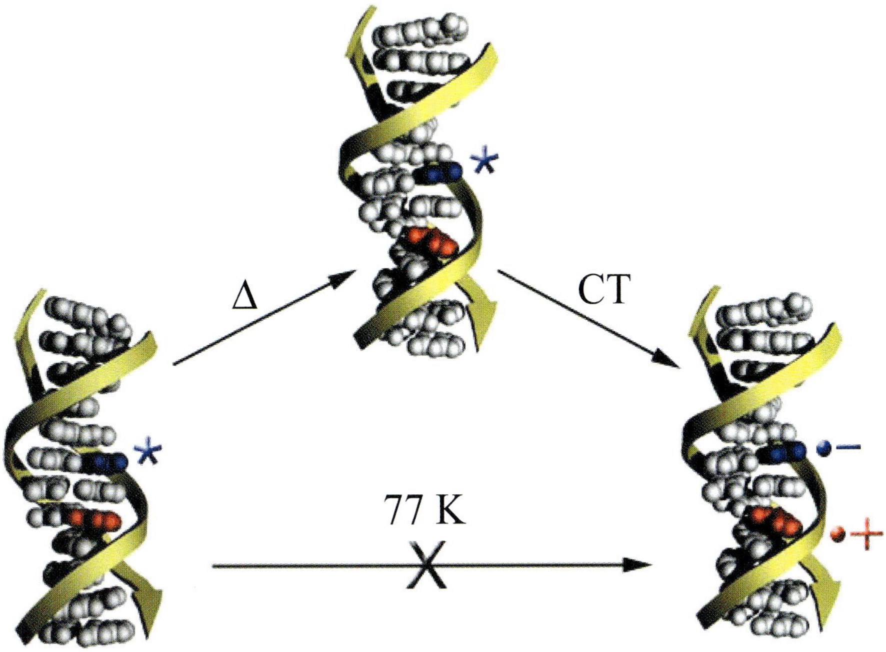

Fig. 2.10 CT in DNA requires base dynamics. In DNA assemblies containing 2-aminopurine (blue) and G (red), CT is observed only at temperatures where base motions are possible.

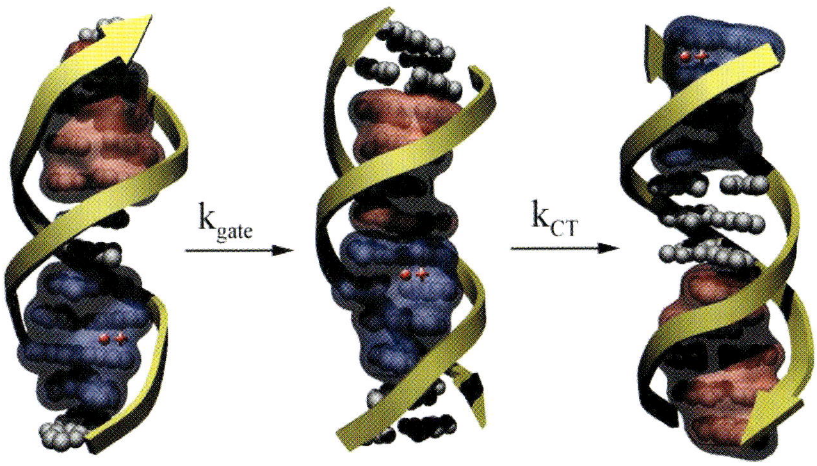

Fig. 2.11 Proposal for CT through DNA through delocalized domains. Injection of a hole into DNA produces a delocalized domain containing the cation radical. Movement of the radical is gated by conformational motions that can produce overlap between the cation radical domain and another well-stacked DNA domain (red). The large domain can then dissolve, again through dynamics, leading to charge transport.

and G is not observed at 77 K; rather than hindering the ability of DNA to transport charge, conformational motion is actually *required*. These observations are completely consistent with our model of conformationally gated CT. Moreover, the lack of DNA-mediated CT at 77 K, even through the shortest bridge, suggests that the static structures adopted upon cooling do not represent optimum CT-active conformations.

2.3.4
Charge Delocalization and Participation of All DNA Bases

Our efforts towards correlating base stack structure and dynamics with DNA CT have also revealed the distance dependence of the yield of base-base CT through relatively long $(A)_n$ bridges (n = 0–9) as a function of temperature [127]. This distance dependence becomes shallower at higher temperatures, as expected for CT that is conformationally gated. More striking, however, was the observation that at all temperatures the yield of CT does not exhibit a simple, monotonic distance dependence but instead displays an oscillatory behavior. This complex distance dependence cannot be fit with either superexchange or hopping models but instead requires a more complete treatment that includes sequence-dependent structure and dynamics. In the context of conformationally gated CT, the distant-dependent yield reflects the influence of the number of bridging bases on the probability of achieving a CT-active conformation. As expected, this probability generally decreases as the number of bases in the CT-active conformation increases. However, in some cases increasing the number of base steps *increases* this probability; for certain numbers of base steps, the probability of achieving a CT-active conformation is enhanced when this number is reached.

To account for both of these observations, we fit the data to an empirical function composed of a monotonic component that decays with the power of n (where n is the number of bases) and a nonmonotonic component [127]. We ascribe the nonmonotonic behavior to coherent motion of groups of DNA bases that leads to the formation of a *domain*. A domain can be described as a transiently extended π-orbital, defined dynamically by DNA sequence, over which charge can delocalize. Transient access to domains is afforded by base dynamics and accounts for conformational gating (Figure 2.11). Based on the period of oscillation seen in our experimental distance dependence, we propose that a domain for our $Ap(A)_nG$ duplexes consists of ca. 4–5 base pairs. During the course of CT through DNA, delocalized domains form and dissolve.

Our experimental approach using Ap* as the oxidant allows us to probe CT through *delocalized* domains. As previously described (*vide infra*), we distinguish CT between Ap* and G by comparing redox-active duplexes to reference duplexes where the G hole donor is replaced by I; the yield of CT evaluated from fluorescence quantum yields is defined as $Fq = 1 - \Phi_G/\Phi_I$. Thus, for CT through intervening DNA bridges, localized injection onto the bridge (i.e., as in an incoherent hopping mechanism) does not contribute to Fq, since this pathway exists in both redox-active and reference duplexes. Instead, Fq describes CT via mechanisms involving superexchange and/or injection of *delocalized* charge. In order for superexchange to account for CT over the experimental distances in the lifetime of Ap* (~5 ps to 10 ns in DNA), β-values $\leq 0.1-0.2$ Å^{-1} would be necessary; such a weak distance dependence suggests that another mechanism may be operative. Moreover, since Ap* may have sufficient energy to oxidize A, a superexchange mechanism need not be invoked. Yet, while our CT reactions likely involve occupation of the DNA bridge, this injection cannot be to a single, electronically isolated base (here, A present in both redox-active and reference duplexes); injection must be sensitive to distant bases (here, I versus G). This sensitivity would arise if the charge is injected into a delocalized domain. Incoherent hopping of localized charge is, therefore, not the only explanation that accounts for a shallow distance dependence of DNA CT. The complex distance dependence we observe excludes localized hopping and can be accounted for through coherent dynamics and charge delocalization in DNA-mediated CT.

That electronic charge can delocalize in duplex DNA is hardly surprising. The delocalized charge and corresponding conductive properties of synthetic π-stacked materials as solids or in solution are well known [129]. More recently, delocalized excitation and charge (holes) over groups of five fluorenes in polyfluorenes that are π-stacked in the solid state and solution were reported [130]. In DNA, exciton delocalization was proposed over 40 years ago [131] and has been the subject of renewed investigations, including a report [131 c] that the number of coherently coupled chromophores is as high as 4–8 for dynamically fluctuating (MD-simulated) duplex DNA. The widely cited gas-phase calculations of Saito and coworkers rationalize guanine oxidation patterns in terms of the distribution of the HOMO asymmetrically delocalized over 5′-GG-3′ doublets [71]. More recently, Renger and Marcus applied a variable-range hopping model to experimental yields of DNA hole transport and found that delocalized bridge states are required to account for the distance dependence [132].

Yet, experimental demonstrations of charge delocalization in DNA, or any D-B-A molecule, are undoubtedly challenging and, not surprisingly, limited. Our earliest investigations that revealed ultrafast DNA-mediated CT over long molecular distances certainly hinted at mechanisms other than superexchange or incoherent hopping [2]. More recent experiments on the sequence sensitivity of long-range guanine oxidation [93] and the yield of CT intermediates [80] indicate that DNA CT cannot be simply described in terms of the motion of holes among guanines. Indeed, the presence of guanines may inhibit transport by modulating stacked conformations [93] or localizing charge [80]. These and other investigations suggest dynamic involvement of the full base pair stack in DNA-mediated CT. We have also shown that charge equilibrates across the duplex on a time scale that is fast compared with trapping at guanine [121]. This does not mean that the charge occupies the entire duplex simultaneously. Instead, our investigations with Ap* indicate that DNA CT is conformationally gated and provide evidence for transient delocalization of holes over domains comprising several bases. Now we are asking more explicit questions regarding the nature of DNA domains. How is charge distributed within a domain? How does this distribution depend on DNA sequence and dynamics?

Is the hole delocalized over the entire base pair, or is it restricted to occupying the presumably lower-energy purines? Can we trap a delocalized hole?

Our first series of investigations aimed at these rather complex issues was based on a very simple concept. Here we exploited again the fast oxidative trap afforded by N_2-cyclopropylamine-substituted bases [59, 60]. Rather than trapping holes on G or A, though, we asked whether, during the course of DNA-mediated CT, holes occupy the higher-energy pyrimidine bases. Could DNA CT induce long-range oxidative damage to pyrimidines? Certainly the answer to this question has profound mechanistic implications for DNA-mediated CT. In particular, the current mechanistic paradigms of superexchange-mediated tunneling and incoherent hopping among low-energy bases (the relative oxidation potentials of the four DNA bases, G, A, C, T, in aqueous solution: ~1.3 V, 1.4 V, 1.6 V, 1.7 V vs. NHE, respectively) [86] predict that long-range oxidation of pyrimidines is not possible.

Our strategy to test for hole occupation of pyrimidines was based on N_4-cyclopropylcytosine (^{CP}C) as the kinetically fast hole trap [61]. As with ^{CP}A [60] and ^{CP}G [59, 60], substitution of the cyclopropyl moiety onto C is expected to have little effect on the base oxidation potential or duplex structure when incorporated into DNA. We observed little change in duplex melting temperature (1 °C) and no change in hypochromism. Our experimental assay employed duplex DNA assemblies containing ^{CP}C positioned remotely from a covalently tethered photooxidant, either $[Rh(phi)_2(bpy')]^{3+}$ (Rh) or an anthraquinone derivative (AQ). Each of these oxidants has sufficient excited-state reduction potential (2.0 V and 1.9 V vs. NHE, respectively) [133, 134] to oxidize each of the four natural DNA bases. Irradiation initiates hole injection and migration through the duplex; the presence of hole density on ^{CP}C can be detected by its decomposition monitored via HPLC. Importantly, ^{CP}C is 4–7 base pairs away from the site of the tethered photooxidant; in order to oxidize ^{CP}C, a hole must first traverse this intervening low-energy bridge.

Remarkably, we observed appreciable oxidative damage at the distant ^{CP}C upon irradiation of either the Rh- or AQ-DNA assemblies. These observations provide direct chemical evidence for the existence of hole density on pyrimidines. Significantly, the damage yield at ^{CP}C was modulated by lower-energy guanine sites on the same or complementary strand. This confirms that the fast trapping reaction probes intrinsic hole density rather than distorting it; the extent of oxidation at ^{CP}C is governed by sequence-dependent variations in the distribution of hole density in duplex DNA. We further observed that the efficiency of trapping at ^{CP}C is equivalent to that at ^{CP}G. Clearly, hole density distribution on the DNA bridge does not reflect the relative energies of the *isolated* bases. Thus, CT through DNA cannot occur by localized, thermally induced charge hopping among only low-energy sites. Instead, all bases participate in DNA CT through charge delocalization within transient, sequence-dependent domains.

These results also shed light on our previous observations of long-range oxidative repair of thymine dimers, another fast trap, via DNA-mediated CT [135–137]. Thymine dimers are the most prevalent photochemical lesions in DNA; while eukaryotic cells excise the thymine dimer, in bacteria the lesion is repaired by photolyase via electron transfer from a reduced flavin cofactor to the cyclobutane dimer [138].

Model studies have demonstrated that the thymine dimer can also be repaired oxidatively (E^0 T<>T ~ 2 V vs. NHE) [139, 140]. We have demonstrated oxidative thymine dimer repair at a distance using potent photooxidants (Rh(III) and NDI) covalently tethered and intercalated within duplex DNA [135–137]. As for ^{CP}C oxidation, efficient thymine dimer repair occurs, despite the fact that the hole must migrate through an intervening DNA bridge possessing much lower energy sites, including guanine doublets whose oxidation should be thermodynamically favored by ~0.7 V. Furthermore, while an intercalated Ru(III) oxidant did not repair the thymine dimer, consistent with the fact that the dimer is not thermodynamically accessible, it did oxidize guanine doublets on *both sides* of the dimer [136]. While difficult to interpret within the framework of models founded solely upon base energetics, these observations are easily rationalized if CT involves dynamic participation of all DNA bases. Significantly, an understanding of the competition between oxidative DNA damage and DNA repair in vivo will certainly provide insight into the consequences and regulation of DNA CT chemistry in biological systems.

2.3.5
Transport of Holes Versus "Excess Electrons"

Our discussion of DNA CT has focused thus far on investigations of the transport of electron holes, processes initiated by DNA-bound oxidants. While investigations of HT have dominated experimental and theoretical studies of DNA-mediated CT, more recently the term “excess electron” transport has been coined to describe the migration of an electron instead of a hole [141]. Although there are distinctions in terms of orbitals involved, we believe that the fundamental parameters described above apply to the migration of charge in general, both electrons and holes. This idea is based on our many investigations of both electron and hole transport in DNA. In fact, the first observation of DNA-mediated CT, made by us over 10 years ago, was rapid, long-range *electron* transport from a photoexcited, intercalated Ru(II) reductant to an electron-accepting Rh(III) complex intercalated over 40 Å away [2].

We frequently employ electrochemical methods to monitor the long-range transport of electrons through DNA films. Here we exploit the ability of DNA to self-assemble on gold surfaces via alkane-thiol linkers. Our DNA-modified surfaces are typically constructed of small (e.g., 15-base-pair) DNA duplexes, tightly packed on gold electrodes through self-assembly in the presence of a high concentration of magnesium ion (Figure 2.12). We have also developed schemes for preparing films of much longer DNA duplexes and utilize much less densely packed films for electrochemical probing of DNA-protein interactions (*vide infra*). Our picture of the DNA-modified surfaces is derived from extensive spectroscopic and biochemical characterization [72, 142, 143], including ^{32}P radioactive labeling to evaluate the surface density, and atomic force microscopy (AFM) studies of the surface morphology. These AFM studies revealed surfaces densely covered with a monolayer 45 Å in thickness, indicating that the DNA duplexes are oriented at an angle of ~45 degrees relative to the gold surface. Furthermore, in AFM studies as a function of potential, we observed that application of a positive potential causes

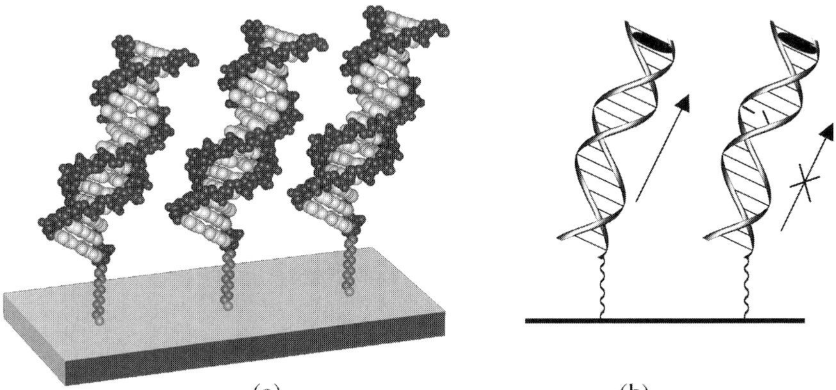

Fig. 2.12 Electron transfer in DNA films. (a) DNA is attached to a gold surface via an alkane-thiol linkage. (b) Electron transfer from the electrode to an intercalator covalently bound at a distant site is facile, but is disrupted when a stacking defect is introduced into the intervening DNA sequence.

the DNA, if not tightly packed, to lie down on the surface, thereby compressing the monolayer (20 Å) [143]. Conversely, application of a negative potential increases the monolayer thickness (50 Å), consistent with the DNA duplexes oriented perpendicular to the surface.

To electrochemically interrogate long-range electron transport through DNA, we attach redox-active intercalators, such as methylene blue (MB), or daunomycin (DM), to these surface-bound DNA duplexes. DNA-mediated transport of an electron from the gold surface to the distally bound intercalator is then monitored by, for instance, cyclic voltammetry or chronocoulometry.

Our investigations of electron transport through these DNA films have been highly complementary to spectroscopic and biochemical assays of hole transport through DNA molecules in solution. Significantly, we found that transport of electrons is governed by the same intrinsic DNA features as the transport of electron holes. DNA-mediated CT proceeds rapidly *through* the π-stacked base pairs in a manner that is extremely sensitive to the structure and dynamics of the base stack, as well as the coupling of the intercalator [144], but largely insensitive to transport distance [73]. CT through the alkane-thiol linker is rate limiting, irrespective of the distance of the redox probe from the gold surface, at least as far as 100 Å [145]! The innate sensitivity of this CT chemistry to π-stack structure and dynamics is the basis of our design and application of novel biosensors for detecting e.g., all single-base mismatches [10], DNA base lesions [10, 146], locked nucleic acids used in antisense technology [147], A, B, and Z DNA polymorphs [148], and DNA-binding proteins [11].

More fundamentally, our well-characterized DNA films have afforded direct probing of the molecular conductivity of B-DNA under physiological conditions using in situ scanning tunneling microscopy (STM) [149]. While some experiments aimed at direct conductivity measurements have characterized DNA as a wide-band-gap semiconductor [150, 151], structural perturbations or poor metal-molecule contacts may inhibit conductivity, particularly given the sensitivity of DNA CT to π-stack structure and flexibility. Instead, we have used STM to characterize the electronic properties of structurally well-defined 15-mer DNA duplexes in aqueous buffer, taking advantage of the innate properties of our thiol-modified DNA films on gold. In particular, since our AFM studies revealed that it is possible to control the orientation of the DNA on the surface (*vide supra*), we used variable potential STM to probe the electronic states of DNA *explicitly* down the helical axis. These experiments revealed that conduction through the duplex is sensitively modulated by the DNA orientation; the local density of states (LDOS) of DNA contribute to electronic communication between the gold and the STM tip only when DNA is in an upright orientation, where effective orbital overlap between the DNA base pairs and the metal electronic states of the tip is possible. However, even for DNA positioned in an upright orientation, conduction could be turned off by the incorporation of a single-base mismatch within the intervening π-stack. Clearly the conductivity arises from π-orbital interactions that are a sensitive function of local electronic structure. These results, completely consistent with our studies of CT between bound molecular probes, confirm that DNA, while not a robust molecular wire, exhibits conductive properties that are a consequence of its unique structure and dynamics.

Significantly, our investigations of the CT properties of these DNA films have further challenged our understanding of how DNA-mediated CT proceeds. In particular, our quandary was how, mechanistically, we could reduce intercalators with reduction potentials ($E_{red} \sim -0.4$–0.6 V vs. NHE) that are significantly below those of the isolated DNA bases (e.g., thymine $E_{red} \sim -1.1$ V vs. NHE) [152]. For an electron to reduce a distally bound intercalator by DNA-mediated CT, it must first traverse an intervening DNA bridge that, based on the reduction potentials of the individual bases, is at least ~0.5 V higher in energy. Noteworthy is the fact that the overall rate of CT in our DNA films is relatively slow (10^2 s^{-1}). Yet, while it is possible that the rate of CT through the DNA base pair stack in DNA films is slower than in solution, for transport over very long distances of 50–100 Å, it appears that what is rate limiting is instead tunneling through the significantly shorter (15 Å) σ-bonded linker [75].

To rationalize CT through these DNA films, we initially hypothesized that the rigidity imposed on the DNA duplexes within the tightly packed films might play a role in stabilizing well-coupled domains. Alternatively, electron transport through DNA films has been discussed in terms of longitudinal polarizability arising from cooperativity effects among neighboring DNA duplexes and their surrounding ionic environment [153]. These hypotheses invoke unique features of the film to account for the CT properties exhibited by these DNA-modified surfaces. Yet we are now seeing these same properties for CT through DNA molecules in solution. In fact, our recent solution experiments demonstrating hole occupancy on pyrimidines during the course of DNA-mediated HT provide a simple mechanistic rationale for the CT observed through DNA films [61]. They also reveal the true complementarity among our spectroscopic, biochemical, and electrochemical investigations, a complementarity possible only because each method interrogates the same intrinsic feature of DNA. The charge transport properties of DNA clearly do not reflect the redox potentials of the individual bases. To understand long-range CT through DNA films, we need not invoke tunneling over unrealistic distances or overcoming a large thermal barrier for occupancy and incoherent hopping. DNA-mediated CT involves dynamic participation of all DNA bases. Charge is transported in delocalized domains onto both the pyrimidines and the purines, in a manner that reflects the sequence- and time-dependent variations in redox potential and electronic coupling within the base pair π-stack.

2.4
A Mechanistic Model for DNA-mediated Charge Transport: Beyond Superexchange and Incoherent Hopping

Through many and varied experiments its has become evident that we cannot accurately describe DNA-mediated CT in terms of incoherent hopping among specific "low-energy" bases. Superexchange tunneling through presumed "high-energy" bases also fails to account for many observations. In particular, our data establish that holes occupy the full DNA bridge, both pyrimidines and purines [61]. Models derived solely upon base energetics furthermore do not provide a reasonable rationale for the sensitivity of CT to sequence-dependent structure and dynamics. An alternative mechanistic description is therefore required.

We have proposed a model where charge migrates through the DNA bridge among delocalized domains (Figure 2.11) [61, 127]. Our data require participation of all bases in DNA-mediated CT; charge may occupy the entire DNA bridge, although not necessarily simultaneously. Instead, these DNA domains can be described as extended π-orbitals formed transiently depending on DNA sequence and dynamics. Spectroscopic investigations of base-base CT with Ap* as a function of bridge length and temperature have provided evidence for a domain size of ca. four base pairs. Data demonstrating comparable efficiency of oxidation at C and G, once their trapping rates are made similar through cyclopropyl substitution, lend direct support for charge delocalization. However, delocalization clearly does not occur over the entire duplex. The transport is partly incoherent since we have also found that CT is gated by base dynamics [126–128]. Thus, through conformational rearrangement of the DNA bases, CT-active domains form and break up transiently, both facilitating and limiting CT. While this model is more challenging to test experimentally, and describe theoretically, than superexchange or localized hopping, as the results presented in this review highlight, DNA-mediated CT requires a more complete treatment.

Our model for CT among delocalized domains is distinct from models that invoke polarons to rationalize experimental observations of long-range guanine oxidation in DNA [1c]. A polaron is a structural distortion created by a charge over which the charge can delocalize; this creates a shallow energy minimum in which the hole becomes self-trapped. This is distinct from a domain, a transiently extended π-orbital over which charge can delocalize. A domain is a sequence-dependent feature of DNA that exists in the absence of the charge. The charge can delocalize over the extended π-orbital without distorting the domain structure and without becoming trapped. Conformational rearrangement of the DNA bases facilitates formation of CT-active domains and movement of charge among domains; DNA CT can therefore be mechanistically described as *conformationally gated transport among delocalized domains*. Conversely, in the phonon-assisted polaron-like hopping model proposed by Schuster and coworkers [1 c], thermally induced fluctuations (phonons) instead provide the energy necessary to overcome the barrier associated with self-trapping of the charge such that it can propagate. Our experimental results provide evidence for delocalized domains, and we need not invoke polarons in a mechanistic description. In fact, several experimental observations are inconsistent with a polaron model, particularly the fact that charge *injection* from Ap* is sensitive to distant bases [127]. This is expected if injection is to a transient extended π-orbital but does not make sense in terms of a polaron that forms only in response to the charge *after* it is injected.

Within our mechanism, it is ultimately the dynamic flexibility of the DNA molecule that makes CT possible and regulates its rate constants, yields, and distance dependences. We propose that of the many conformations a DNA molecule can adopt, only those with particular arrangements of the DNA bases will afford CT-active domains, and that base dynamics, necessary to achieve these conformations, gate CT. Alternatively, it has been suggested that it is not the conformation of the DNA bases that is important but rather the configuration of the hydrated counterions; dynamic fluctuations in ionic configuration gate CT [154]. This model is based on a combined electronic-structure, molecular-dynamics investigation of the spatial distribution of the hole in DNA as a function of ionic configuration. In the simulation, the DNA was held in a fixed conformation and only the hydrated counterions were free to move; interestingly, the results found hole density not only on the DNA bases but also on the sugar-phosphate backbone and the water molecules. Furthermore, only one experimental test of the ion-gated mechanism was made. In this experiment a segment of the negatively charged phosphate backbone between two GG sites, distal and proximal to an end-capped anthraquinone photooxidant, was replaced with neutral methyl phosphonates; this substitution was found to reduce the yield of permanent G oxidation at the distal site [154]. However, since backbone substitution with methyl phosphonate has been shown to bend the DNA up to ~20 degrees towards the neutral face [155], these experiments do not address ionic configuration.

While it is undoubtedly true that hydrated counterions are key to the structure and dynamics of DNA, it is the electronic structure *within* the DNA π-stack that governs DNA-mediated CT. Small changes in base conformation are expected to have a much greater impact on this electronic structure [122] than comparable changes in ionic configuration. In fact, our model for gating through base dynamics emerges from several pieces of experimental evidence [127] and is supported by theory [156]. First, using Ap* as a dual reporter of structural dynamics and base-base CT, we have shown that temperature-induced changes in CT yield are coincident specifically with changes in base dynamics. Second, the influence of temperature on the yield of CT is strongly dependent on the number of bases in the intervening bridge. This is expected for conformational gating mediated through base dynamics, but not for gating through ionic fluctuations. Third, the CT yield and temperature dependence are independent of the identity of the monovalent counterion (Na^+ or K^+).

Unlike models derived from the redox energies of isolated DNA bases, our model originates in DNA's inherent structural dynamics. Consequently, it can account for varied experimental observations of DNA CT both in designed assemblies and in more complex biological milieus. The importance of conformational rearrangements and gating to biological molecules is certainly not a new concept. In fact, conformational substates are a reigning paradigm for protein function [157]. Yet similar models for DNA CT are surprisingly limited [156]. Perhaps this is due to analogies made between the DNA base stack and rigid, solid-state π-stacked materials, or to analogies made between DNA CT and intraprotein ET, which tends to exhibit a well-described dependence of kinetics on distance and driving force [16]. But, with chains of redox participants separated by ~ ≤ 14 Å and an intervening bridge significantly higher in energy, superexchange-mediated tunneling easily transports electrons within proteins on a time scale that is fast compared to rate-limiting catalysis. Consequently it has been suggested that intraprotein ET is quite resistive to both conformational fluctuations and mutational variations within the ET pathway [158]. But biological roles of DNA CT must certainly be distinct from those of redox-active proteins. While Nature apparently designed redox proteins for robust, short-range electronic communication that would not be easily modulated by dynamics or mutations, it appears that CT through DNA may instead be optimized for electronic communication over a much longer range, but with a far greater sensitivity to the intervening bridge: a "fragile wire" to be short-circuited in the presence of a mutation or perturbation in its pathway.

2.5
DNA-mediated Charge Transport in Biology

Our mechanistic investigations have revealed how DNA exploits its intrinsic flexibility to accomplish long-range charge transport that is precisely regulated through base stack structure and dynamics. They also underscore the distinct mechanistic approaches to CT in DNA versus typical redox proteins. These findings certainly provide clues regarding the physiological roles of CT through DNA. How does Nature harness the inherent CT properties of the DNA π-stack? We are now focusing on a variety of experiments to address this fundamental question.

Unlike the naked DNA typical of our in vitro assemblies, in eukaryotic cells, DNA is packaged in nucleosome core particles; ~150 base pairs of DNA are coiled around an octamer of histone proteins to which the DNA is associated by nonspecific electrostatic interactions. Would CT proceed through DNA within the nucleosome core particle? Would the restricted motion and overall bending of the DNA within the nucleosome disrupt long-range CT? We investigated DNA-mediated CT through the nucleosome core particle using photoexcited Rh(III) intercalators, either noncovalently bound or covalently tethered to the 146-base-pair DNA duplexes packaged into nucleosomes [12]. Remarkably, although the histone proteins inhibited intercalation of the rhodium complex within the core particle, they did not prevent oxidative CT through the DNA over distances up to 75 Å. Although histone binding did not influence the extent or pattern of oxidative damage, guanine oxidation was not detected at more distance sites even in the absence of the histone proteins; likely, the bent DNA structure used to generate consistent nucleosome phasing interfered with CT. Significantly, then, although packing of the DNA in nucleosomes may provide protection from solution-mediated damage, it does not protect against oxidative damage via DNA-mediated CT.

To further explore DNA-mediated CT in the physiological realm, we probed CT within cell nuclei. In the first experiments we irradiated Rh(III) intercalators that had been incubated with HeLa nuclei [13]. The DNA was then isolated and treated with base-excision repair enzymes to reveal oxidative damage as strand breaks, and the damaged DNA was amplified using the ligation-mediated polymerase chain reaction. Oxidative damage was observed preferentially at the 5′-G of 5′-GGG-3′, 5′-GG-3′, and 5′-GA-3′ sites, indicative of DNA-mediated CT chemistry. Importantly, oxidation was detected at sites that, due to protein binding, are inaccessible to the Rh(III) intercalator, demonstrating DNA-mediated CT from a distance on transcriptionally active DNA within the cell nucleus. These exciting observations have propelled our studies of in vivo DNA CT to the forefront.

Given that long-range DNA-mediated CT occurs within the cellular milieu, what might be the sources of electrons and holes in vivo? Our biochemical and electrochemical assays have demonstrated that DNA-binding proteins can influence π-stack structure and dynamics and therefore CT, but can proteins be active participants in CT reactions involving DNA? For instance, can electrons and holes be transferred between DNA and proteins, and can these reactions be triggered at a distance?

In our first experiments, using the flash-quench technique, we observed long-range oxidation of both tyrosine and tryptophan when Lys-X-Lys (X = tyrosine or tryptophan) tripeptides were bound to DNA [159, 160]. These results established direct participation of the DNA-bound peptides in long-range CT through DNA. We then examined both the transient intermediates and the yields of oxidative damage induced upon flash-quench generation of a Ru(III) oxidant intercalated within DNA containing the M.*Hha* I binding site [161]. Upon irradiation of an assembly with the mutant M.*Hha* I (Q237W) bound 14 base pairs away from the Ru(III), significant oxidation was detected at the guanine located 3′ to the site of tryptophan intercalation. Significantly, no such damage was evident upon binding of the wild-type enzyme that instead inserts a glutamine into the DNA base stack. In transient absorption experiments, both the guanine and tryptophan ($E^{0/+} \sim 1$ V vs. NHE) radicals were detected as intermediates. Thus, we have directly observed long-range CT through DNA to the inserted tryptophan of a bound protein and to the adjacent 5′-guanine yielding permanent oxidative damage. Interestingly, the yield of oxidation decreased only slightly with increasing distance (24–51 Å) between the M.*Hha* I binding site and the intercalated Ru(III) oxidant, while the rate constant for radical formation was unchanged ($>10^6\ s^{-1}$) irrespective of distance. Consequently, CT through DNA over this 50-Å distance regime is not rate limiting, consistent with analogous investigations employing the artificial base methylindole directly incorporated within the DNA base stack [56].

In our studies of DNA-protein CT, we were naturally interested in probing for electron transfer from a protein to oxidatively damaged DNA. We first examined CT from ferrocytochrome C to a guanine radical in DNA [162]. The flash-quench technique was used to generate the potent ground-state oxidant, $[Ru(phen)_2dppz]^{3+}$, intercalated within poly(dG-dC). Transient absorption measurements revealed rapid oxidation of guanine by the Ru(III) intercalator, followed by slower CT from the ferrocytochrome C to the resultant guanine radical. Thus, repair of oxidative damage on DNA by DNA-protein CT is indeed possible.

Interestingly, however, many DNA-repair proteins do not use ET chemistry to mend damaged bases. While we have a reasonable understanding of catalytic mechanisms of the repair reactions, we know little about the mechanism of damage location by DNA-repair proteins. This is a particularly daunting task. Maintaining the integrity of the genome is obviously essential for survival, yet it requires proteins that are generally low in copy number to locate single-base lesions, often structurally similar to natural bases, within our three-billion-base-pair genome. How does Nature accomplish this fundamental task? As we devised schemes to detect mismatches in vitro by exploiting the exquisite sensitivity of DNA-mediated CT to π-stack structure and dynamics, we naturally wondered whether Nature might do the same [163].

We also noticed that many base-excision repair enzymes possess $[4Fe4S]^{2+}$ clusters that have no apparent catalytic or structural role. Could these clusters instead be involved in damage recognition? To begin addressing that question we have conducted electrochemical investigations of MutY, an FeS-containing base-excision repair enzyme from *E. coli*, bound to our DNA-modified gold electrodes [14]. Although the [4Fe4S] cluster of MutY is electrochemically inert in the absence of DNA, we observed oxidation of the cluster when bound to DNA in a DNA-mediated reaction; binding to DNA induces a change in the $[4Fe4S]^{3+/2+}$ potential, activating oxidation.

Given this DNA-mediated CT involving the FeS cluster, we proposed that MutY, and other DNA-repair proteins, might exploit long-range CT to locate mismatches and lesions in DNA [14, 164] (Figure 2.13). In this model, nonspecific binding of MutY to DNA shifts the redox potential of the cluster, promoting its oxidation and thereby injecting an electron into the DNA base pair stack. Through DNA-mediated CT, this electron could reduce a distally bound MutY, or another repair protein, provided there were no intervening lesions perturbing the π-stack. Reduction of the distally bound protein lowers the binding affinity, facilitating dissociation of the reduced protein from the DNA. In this way, a relatively large region of the genome could be rapidly scanned for lesions by relatively few proteins. The model resembles the way telephone repairmen might look for a mistake in the

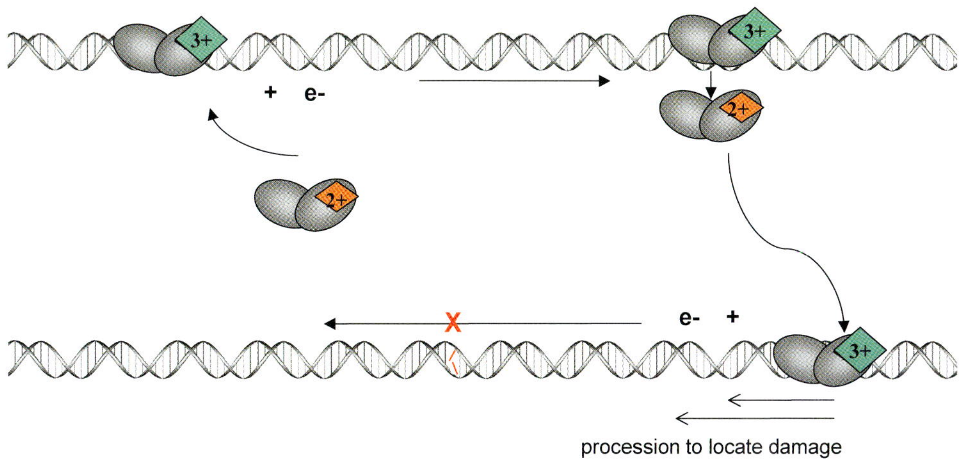

Fig. 2.13 A CT scan for detecting DNA damage. Binding of MutY (or other FeS DNA-repair proteins) leads to activation of the FeS cluster for electron transfer, and a CT event to a distant DNA-bound protein represents a successful scan of the DNA (top). Once reduced to the 2+ cluster, the protein dissociates from the DNA. In the presence of a defect in the sequence intervening the proteins, the CT event is blocked, the oxidized 3+ protein does not dissociate, and, instead, the repair protein processes on a slower time scale toward the damage (bottom).

line. Like the repairmen, the proteins leave upon getting the signal and check somewhere else. Should a lesion exist in the region near where the protein binds, the perturbation in π-stack structure and dynamics would inhibit the DNA-mediated CT. Consequently, the distally bound protein would remain associated with the DNA, processively diffusing to the mismatch site on a slower time scale. This hypothesis presents a fascinating, and potentially significant, physiological role for DNA-mediated CT that fully exploits the mechanistic features uncovered in our in vitro investigations.

2.6 Conclusions and Outlook

The remarkable fact that electrons and holes can shuttle through the DNA π-stack gives us new perspective on this celebrated molecule of life. Through a decade of creative spectroscopic, biochemical, and electrochemical investigations, the mystery of how DNA can mediate CT has evolved into a coherent mechanistic model. While models derived upon base energetics were a logical first step, they are surrogates adopted to, but not designed for, DNA. To describe DNA CT, we must begin with the sequence-dependent structure and dynamics that are intrinsic to DNA and essential for DNA function. As we have discovered, charge migration through DNA involves delocalization within transient, structure-dependent domains that dynamically involve all of the DNA bases. Charge distribution and transport are tightly controlled by conformational dynamics of the base pair stack. With this mechanistic description, the sensitivity of DNA CT to base pair structure can be reconciled and exploited and its physiological roles can be elucidated. Charge transport through DNA is clearly distinct mechanistically from proteins, where tunneling predominates and distance and energetics are indeed decisive factors. These distinctions reflect different biological roles of CT in DNA as compared to proteins.

Investigations in vitro suggest that DNA-mediated CT may be important in long-range cellular DNA damage and repair, and we are continuing to probe the patterns of oxidative damage in various regions of the genome. We have also proposed a physiological role for DNA CT that explicitly exploits its dependence on base stack structure and dynamics. Here, DNA-repair proteins may harness DNA CT for rapid, sensitive detection of DNA lesions. Our current experiments are designed to confirm this proposal and to expose other areas of biology where DNA CT may play a role. Just as we aim to exploit DNA CT in molecular electronics and biosensors, we would hardly be surprised to discover that Nature relies on DNA's unique CT properties to accomplish long-range communication, signaling, and other cellular tasks.

Acknowledgments

We are grateful to our many coworkers and collaborators cited throughout this review. We thank, in particular, E. D. A. Stemp for continuing to participate in this remarkable journey of discovery. We also thank NIH and NFCR for their financial support.

References

1 For recent reviews see: (a) *Topics in Current Chemistry: Long-range Electron Transfer in DNA:1*, ed. G. B. Schuster, Springer, **2004**, *236*. (b) S. Delaney, J. K. Barton, *J. Org. Chem.* **2003**, *68*, 6475–6483. c) G. B. Schuster, *Acc. Chem. Res.* **2000**, *33*, 253–260. (d) Giese, B. *Curr. Op. Chem. Biol.* **2003**, *6*, 612–618. e) F. D. Lewis, R. L. Letsinger, M. R. Wasielewski, *Acc. Chem. Res.* **2001**, *34*, 159–170.

2 C. J. Murphy, M. R. Arkin, Y. Jenkins, N. D. Ghatlia, S. H. Bossmann, N. J. Turro, J. K. Barton, *Science* **1993**, *262*, 1025–1028.

3 C. Z. Wan, T. Fiebig, S. O. Kelley, C. R. Treadway, J. K. Barton, A. H. Zewail, *Proc. Natl. Acad. Sci. U.S.A.* **1999**, *96*, 6014–6019.

4 C. Z. Wan, T. Fiebig, O. Schiemann, J. K. Barton, A. H. Zewail, *Proc. Natl. Acad. Sci. USA* **2000**, *97*, 14052–14055.

5 F. D. Lewis. X. Zuo, J. Liu, R. T. Hayes, M. R. Wasielewski, *J. Am. Chem. Soc.* **2002**, *124*, 4568–4569.

6 M. E. Núñez, D. B. Hall, J. K. Barton, *Chem. Biol.* **1999**, *6*, 85–97.
7 P. T. Henderson, D. Jones, G. Hampikian, Y. Z. Kan, G. B. Schuster, *Proc. Natl. Acad. Sci. USA* **1999**, *96*, 8353–8358.
8 D. B. Hall, R. E. Holmlin, J. K. Barton, *Nature* **1996**, *382*, 731–735.
9 P. J. Dandliker, R. E. Holmlin, J. K. Barton, *Science* **1997**, *275*, 1465–1468.
10 E. M. Boon, D. M. Ceres, T. G. Drummond, M. G. Hill, J. K. Barton, *Nature Biotechnology* **2000**, *18* , 1096–1100.
11 E. M. Boon, J. E. Salas, J. K. Barton, *Nature Biotechnology* **2002**, *20*, 282–286.
12 M. E. Núñez, K. T. Noyes, J. K. Barton, *Chem. Biol.* **2002**, *9*, 403–415.
13 M. E. Núñez, G. P. Holmquist, J. K. Barton, *Biochemistry* **2001**, *40*, 12465–12471.
14 E. M. Boon, A. L. Livingston, N. H. Chmiel, S.S. David, J. K. Barton, *Proc. Natl. Acad. Sci. USA* **2003**, *100*, 12543–12547; **2004**, *101*, 4718.
15 See e. g., a) M. Bixon, B. Giese, S. Wessely, T. Langenbacher, M. E. Michel-Beyerle, J. Jortner, *Proc. Natl. Acad. Sci. USA* **1999**, *96*, 11713–11716. b) Y. A. Berlin, A. L. Burin, M. A. Ratner, *Chem. Phys.* **2002**, *275*, 61–74. c) F. D. Lewis, J. Q. Liu, X. B. Zuo, R. T. Hayes, M. R. Wasielewski, *J. Am. Chem. Soc.* **2003**, *125*, 4850–4661. d) B. Giese, J. Amaudrut, A. K. Kohler, M. Spormann, S. Wessely, *Nature* **2001**, *412*, 318–320.
16 H. B. Gray, J. R. Winkler, *Annu. Rev. Biochem.* **1996**, *65*, 537–561.
17 W, Sanger, *Principles of nucleic acid structure*, Cantor CR (ed), Springer-Verlag. **1984.**
18 J. D. Watson, F. H. C. Crick *Nature* **1953**, *171*, 737–738.
19 A. Szent-Györgi, *Proc. Natl. Acad. Sci. USA.* **1960**, *46*, 1444–1449.
20 D. D. Eley, D. I. Spivey, *Trans. Faraday Soc.* **1962**, *58*, 411–415.
21 see e. g., (a) T. M. Nordlund, S. Andersson, L. Nilsson, R. Rigler, A. Gräslund, L. W. McLaughlin, *Biochemistry* **1989**, *28*, 9095–9013. (b) C. R. Guest, R. A. Hochstrasser, L. C. Sowers, D. P. Millar, *Biochemistry* **1991**, *30*, 3271–3279. (c) E. B. Brauns, M. L. Madaras, R. S. Coleman, C. J. Murphy, M. A. Berg, *J. Am. Chem. Soc.* **1999**, *121*, 11644–11649. (d) E. B. Brauns, C. J. Murphy, M. A. Berg, *J. Am. Chem. Soc.* **1998**, *120*, 24492456-. (e) Z. Liang, J. H. Freed, R. S. Keyes, A. M. Bobst, *J. Phys. Chem. B* **2000**, *104*, 5372–5381. (f) S. Georghiou, T. D. Bradrick, A. Philippetis, J. M. Beechem, *Biophys. J.* **1996**, *70*, 1909–1922. (g) E. Giudice, R. Lavery, *Acc. Chem. Res.* **2002**, *35*, 350–357.
22 M. E. Núñez, J. K. Barton, *Curr. Opin. Chem. Biol.* **2000**, *4*, 199–206.
23 K. E. Erkkila, D. T. Odom, J. K. Barton, *Chem. Rev.* **1999**, *99*, 2777–2795.
24 R. E. Holmlin, P. J. Dandliker, J. K. Barton, *Bioconjugate Chem.* **1999**, *10*, 1122–1130.
25 A. Sitlani, E. C. Long, A. M. Pyle, J. K. Barton, *J. Am. Chem. Soc.* **1992**, *114*, 2303–2312.
26 A. Sitlani, J. K. Barton, *Biochemistry* **1994**, *33*, 12100–12108.
27 S.S. David, J. K. Barton, *J. Am. Chem. Soc.* **1998**, *115*, 2984–2985.
28 B. P. Hudson, J. K. Barton, *J. Am. Chem. Soc.* **1998**, *120*, 6877–6888.

81 V. Shafirovich, A. Dourandin, W. D. Huang, N. P. Luneva, N. E. Geacintov, *Phys. Chem. Chem. Phys.* **2000**, *2*, 4399–4408.
82 K. Fukui, K. Tanaka, M. Fujitsuka, A. Watanabe, O. Ito, *J. Photochem. Photobiol. B.* **1999**, *50*, 18–27. (b) D. Ly, L. Sanii, G. B. Schuster, *J. Am. Chem. Soc.* **1999**, *121*, 9400–9410. (c) Y. T. Long, C. Z. Li, T. C. Sutherland, M. Chahma, J. S. Lee, H. B. Kraatz, *J. Am. Chem. Soc.* **2003**, *125*, 8274–8725. (d) C. Haas, K. Kräling, M. Cichon, N. Rahe, T. Carell, *Angew. Chem. Int. Ed.* **2004**, *43*, 1842–1844.
83 (a) S. Hess, M. Gotz, W. B. Davis, M. E. Michel-Beyerle, *J. Am. Chem. Soc.* **2001**, *123*, 10046–100055. (b) W. B. Davis, I. Naydenova, R. Haselsberger, A. Ogrodnik, B. Giese, M. E. Michel-Beyerle, *Angew. Chem. Int. Ed.* **2000**, *39*, 3649–3652.
84 K. Nakatani, C. Dohno, I. Saito, *J. Am. Chem. Soc.* **1999**, *121*, 10854–10855.
85 M. A. O'Neill, J. K. Barton, *Proc. Natl. Acad. Sci. USA* **2002**, *99*, 16543–16550.
86 M. A. O'Neill, H.C. Becker, C. Wan, J. K. Barton, A. H. Zewail, unpublished results
87 S. Steenken, S.V. Jovanovic, *J. Am. Chem. Soc.* **1997**, *119*, 617–618.
88 D. B. Hall, J. K. Barton, *J. Am. Chem. Soc.* **1997**, *119*, 5045–5046.
89 F. Prat, K. N. Houk, C. S. Foote, *J. Am. Chem. Soc.* **1998**, *120*, 845–846.
90 P. T. Henderson, D. Jones, G. Hampikian, Y. Z. Kan, G. B. Schuster,
91 M. R. Arkin, E. D. A. Stemp, S.C. Pulver, J. K. Barton, *Chem. Biol.* **1997**, *4*, 389–400.
92 D. B. Hall, S. O. Kelley, J. K. Barton, *Biochemistry* **1998**, *37*, 15933–15940.
93 T.T. Williams, D. T. Odom, J. K. Barton, *J. Am. Chem. Soc.* **2000**, *122*, 9048–9049.
94 E. Meggers, M. E. Michel-Beyerle, B. Giese, *J. Am. Chem. Soc.* **1998**, *120*, 12950–12955.
95 M. Bixon, J. Jortner, *J. Phys. Chem. B* **2000**, *104*, 3906–3913.
96 J. Jortner, M. Bixon, T. Langenbacher, M. E. Michel-Beyerle, *Proc. Natl. Acad. Sci. USA.* **1998**, *95*, 12759–12765.
97 J. G. Nadeau, D. M. Crothers, *Proc. Natl. Acad. Sci. USA.* **1989**, *86*, 2622–2626.
98 R. E. Dickerson, *Nucleic Acids Res.* **1998**, *26*, 1906–1926.
99 T. Brown, W. N. Hunter, G. Kneale, O. Kennard, *Proc. Natl. Acad. Sci. USA.* **1986**, *83*, 2402–2406.
100 W. N. Hunter, T. Brown, O. Kennard, *Nucleic Acids Res.* **1987**, *15*, 6589–6606.
101 H. T. Allawi, J. SantaLucia Jr, *Nucleic Acids Res.* **1998**, *26*, 4925–4934.
102 P. K. Bhattacharya, J. K. Barton, *J. Am. Chem. Soc.* **2001**, *123*, 8649–8658.
103 B. Giese, S. Wessely, *Angew. Chem. Int. Ed.* 2000, 39, 3490–3491.
104 M. E. Núñez, K. T. Noyes, D.A. Gianolio, L. W. McLaughlin, J. K. Barton, *Biochemistry* **2000**, *39*, 6190–6199.
105 D. T. Odom, J. K. Barton, *Biochemistry* **2001**, *40*, 8727–8737.

106 D. T. Odom, E. D. Dill, J. K. Barton, *Nucleic Acids Res.* **2001**, *29*, 2026–2033.

107 D. T. Odom, E. D. Dill, J. K. Barton, *Chem. Biol.* **2000**, *7*, 475–481.

108 S. Delaney, J. K. Barton *Biochemistry* **2003**, *42*, 14159–14165.

109 J. R. Williamson, M. K. Raghuraman, T. R. Cech, *Cell* **1989**, *59*, 871–880.

110 (a) T. von Zglinicki, *Trends Biochem. Sci.* **2002**, *27*, 339–344. (b) A. Heller, *Faraday Discuss.* **2000**, 1013.

111 S. R. Rajski, S. Kumar, R. J. Roberts, J. K. Barton, *J. Am. Chem. Soc.* **1999**, *121*, 5615–5616.

112 S. R. Rajski, J. K. Barton, *Biochemistry* **2001**, *40*, 5556–5564.

113 E. M. Boon, M. A. Pope, S. D. Williams, S.S. David, J. K. Barton, *Biochemistry* **2002**, *41*, 8464–8470.

114 G. D. Reid, D. J. Whittaker, M. A. Day, D. A. Turton, V. Kasyer, J. M. Kelly, G. S. Beddard *J. Am. Chem. Soc.* **2002**, 124, 5518–5527.

115 F. D. Lewis, J. Liu, W. Weigel, W. Rettig, I. V. Kurnikov, D. N. Beratan, *Proc. Natl. Acad. Sci. USA* **2002**, *99*, 12536–12541.

116 B. Giese, *Ann. Rev. Biochem.*, **2002**, *71*, 51–70.

117 F. D. Lewis, T. Wu, X. Liu, R. L. Letsinger, S. R. Greenfield, S. E. Miller, M. R. Wasielewski, *J. Am. Chem. Soc.* **2000**, *122*, 2889–2902.

118 C. Dohno, E. D. A. Stemp, J. K. Barton, *J. Am. Chem. Soc.* **2003**, 125, 9586–9587.

119 T. T. Williams, C. Dohno, E. D. A. Stemp, J. K. Barton, *J. Am. Chem. Soc.* **2004**, 126, 8148–8158.

120 K. Kawai, T. Takada, T. Nagai, X. Cai, A. Sugimoto, M. Fujitsuka, T. Majima, *J. Am. Chem. Soc.* **2003**, 125, 16198–16199.

121 S. Delaney, J. Yoo, E. D. A. Stemp, J. K. Barton, *Proc. Natl. Acad. Sci. USA.* **2004**, *101*, 10511–10516.

122 (a) A. A. Voityuk, K. Siriwong, N. Rösch, *Phys. Chem. Chem. Phys.* **2001**, *3*, 5421–5425. (b) A. Troisi, G. Orlandi, *J. Phys. Chem. B* **2002**, *106*, 2093–2101.

123 A. A. Voityuk, K. Siriwong, N. Rösch, *Angew. Chem. Int. Ed.* **2004**, *43*, 624–627.

124 T. Fiebig, C. Z. Wan, S. O. Kelley, J. K. Barton, A. H. Zewail, *Proc. Natl. Acad. Sci. USA.* **1999**, *96*, 1187–1192.

125 Dynamic disorder has been suggested to decrease electronic coupling, e.g., (a) F. C. Grozema, L. D. A. Siebbeles, Y. A. Berlin, M. A. Ratner, *Chem. Phys. Chem.* **2002**, *6*, 536–539, and to localize charge, e.g., (b) M. Hjort, S. Stafström, *Phys. Rev. Lett.* **2001**, *87*, 228101.

126 M. A. O'Neill, H. C. Becker, C. Wan, J. K. Barton, A. H. Zewail, *Angew. Chem. Int. Ed.* **2003**, *42*, 5896–5900.

127 M. A. O'Neill, J. K. Barton, *J. Am. Chem. Soc.* **2004**, *126*, 11471–11483.

128 M. A. O'Neill, J. K. Barton, *J. Am. Chem. Soc.* **2004**, *126*, in press.

129 See e.g., (a) T. J. Marks, *Science* **1985**, *227*, 881–889. (b) L. L. Miller, K. R. Mann, *Acc. Chem. Res.* **1996**, *29*, 417–423. (c) D. D. Graf, R. G. Duan, J. P. Campbell, L. L. Miller, K. R. Mann, *J. Am. Chem. Soc.* **1997**, *119*, 5888–5899.

130 (a) T. Nakano, T. Yade, *J. Am. Chem. Soc.* **2003**, *125*, 15474–15484. (b) T. Nakano, K. Takewaki, T. Yade, Y. Okamato, *J. Am. Chem. Soc.* **2001**, *123*, 9182–9183. (c) R. Rathore, S. H. Abdelwahed, I. A. Guzei, *J. Am. Chem. Soc.* **2003**, *125*, 8712–8713.
131 (a) I. Tinoco Jr., *J. Am. Chem. Soc.* **1960**, *82*, 4785–4790. (b) B. Bouvier, D. Gustavsson, D. Markovitsi, P. Millié, *Chem. Phys.* **2002**, *275*, 75–92. (c) B. Bouvier, J. P. Dognon, R. Lavery, D. Markovitsi, P. Millié, D. Onidas, K. Zakrzewska, *J. Phys. Chem. B* **2003**, *107*, 13512–13522. (d) M. Rist, H. A. Wagenknecht, T. Fiebig, *Chem. Phys. Chem.* **2002**, *3*, 704–708.
132 T. Renger, R. A. Marcus, *J. Phys. Chem. A* **2003**, *107*, 8404–8419.
133 C. Turro, A. Evenzahav, S. H. Bossmann, J. K. Barton, N. J. Turro, *Inorg. Chim. Acta.* **1996**, *243*, 101–108.
134 B. Armitage, C. Yu, C. Devadoss, G. B. Schuster, *J. Am. Chem. Soc.* **1994**, *116*, 9847–9859.
135 P. J. Dandliker, R. E. Holmlin, J. K. Barton, *Science* **1997**, *275*, 1465.
136 P. J. Dandliker, M. E. Núñez, J. K. Barton, *Biochemistry* **1998**, *37*, 6491–6502.
137 D. A. Vicic, D. T. Odom, M. E. Núñez, D. A. Gianolio, L. W. McLaughlin, J. K. Barton, *J. Am. Chem. Soc.* **2000**, *122*, 8603–8611.
138 A. Sancar, *Annu. Rev. Biochem.* **1996**, *65*, 43–81.
139 J. R. Jacobsen, A. G. Cochran, J. C. Stephans, D. S. King, P. G. Schultz, *J. Am. Chem. Soc.* **1995**, *117*, 5453–5461.
140 T. Young, R. Nieman, S. D. Rose, *Photochem. Photobiol.* **1990**, *52*, 661–668.
141 See e.g., (a) (b) Z. L. Cai, X. F. Li, M. D. Sevilla *J. Phys. Chem. B* **2002**, *106*, 2755–2762. (b) C. Behrens, L.T. Burgdorf, A. Schwogler, T. Carell, *Angew. Chem. Int. Ed.* **2002**, *41*, 1763–1766.
143 (a) M. Sam, E. M. Boon, J. K. Barton, M. G. Hill, E. M. Spain, *Langmuir* **2001**, *17*, 5727–5730. (b) S. O. Kelley, J. K. Barton, N. M. Jackson, L. D. McPherson, A. B. Potter, E. M. Spain, M. J. Allen, M. G. Hill, *Langmuir* **1998**, *14*, 6781–6784.
142 S. O. Kelley, E. M. Boon, J. K. Barton, N. M. Jackson, M. G. Hill, *Nucleic Acids Res.* **1999**, *27*, 4830–4837.
144 E. M. Boon, N. M. Jackson, M. A. Wightman, S. O. Kelley, M. G. Hill, J. K. Barton *J. Phys. Chem. B* **2003**, *106*, 2755–2762.
145 T. Liu, J. K. Barton, unpublished results.
146 A. Boal, J. K. Barton, *Bioconjugate Chem.* **2005**, in press.
147 E. M. Boon, J. K. Barton, P. I. Pradeepkumar, J. Isaksson, C. Petit, J. Chattopadhyaya, *Angew. Chem.Int. Ed.* **2002**, *41*, 3402–3405.
148 E. M. Boon, J. K. Barton, *Bonconjugate Chem.* **2003**, *14*, 1140–1147.
149 D. M. Ceres, J. K. Barton, *J. Am. Chem. Soc.* 2003, *125*, 14064–14065.
150 D. Porath, A. Bezryadin, S. D. Vries, C. Dekker, *Nature* **2000**, *403*, 635–638.
151 H. Fink, C. Schonenberger, *Nature* **1999**, *398*, 407–410.

152 S. Steenken, J. P. Telo, H. M. Novais, L. P. Candeias, *J. Am. Chem. Soc.* **1992**, *114*, 4701–4709.
153 G. Hartwich, D. J. Caruana, T. de Lumley-Woodyear, Y. B. Wu, C. N. Campbell, A. Heller, *J. Am. Chem. Soc.* **1999**, *121*, 10803–10812.
154 R. N. Barnett, C. L. Cleveland, A. Joy, U. Landman, G. B. Schuster, *Science* **2001**, *294*, 567–571.
155 (a) J. K. Strauss, L. J. Maher, *Science* **1994**, *266*, 1829–1834. (b) J. K. Strauss-Soukup, P. D. Rodrigues, L. J. Maher, *Biophys. Chem.* **1998**, *72*, 297–306.
156 (a) M.R. D'Orsogna, J. Rudnick, *Phys. Rev. E.* **2002**, *66*, artno041804. (b) R. Bruinsma, G. Gruner, M. R. D'Orsogna, J. Rudnick, J. *Phys. Rev. Lett.* **2000**, *85*, 4393–4396.
157 H. Frauenfelder, S. G. Sligar, P. G.Wolynes, *Science* **1991**, *254*, 1598–1603.
158 C. C. Page, C. C. Moser, P. L. Dutton, *Current Opin. Chem. Biol.* **2003**, *7*, 551–556.
159 H. A. Wagenknecht, E. D. A. Stemp, J. K. Barton, *Biochemistry* **2000**, *39*, 5483–5491.
160 H. A. Wagenknecht, E. D. A. Stemp, J. K. Barton, *J. Am. Chem. Soc.* **2000**, *122*, 1–7.
161 H. A. Wagenknecht, S. R. Rajski, M. Pascaly, E. D. A. Stemp, J. K. Barton, *J. Am. Chem. Soc.* **2001**, *123*, 4400–4407.
162 E. D. A. Stemp, J. K. Barton, *Inorg. Chem.* **2000**, *39*, 3868–3874.
163 S. R. Rajski, B. A. Jackson, J. K. Barton *Mutation Res.* **2000**, *447*, 49–72.
164 E. Yavin, A. K. Boal, E. D. A. Stemp, E. M. Boon, A. L. Livingston, V. L. O'Shea, S. S. David, J. K. Barton, *Proc. Natl. Acad. Sci. USA* **2005**, in press.

3
Excess Electron Transfer in DNA Probed with Flavin- and Thymine Dimer-modified Oligonucleotides

Thomas Carell and Martin von Meltzer

3.1
Introduction

Today strong efforts are under way to create DNA-inspired, electronically active materials with self-organizing properties [1–4]. The hope is that such a novel material may self-assemble into complex conductive nano-wire networks [5, 6], which may have fascinating technological potential. In this context, the question of how electrons travel through DNA is of fundamental importance [7]. Investigations into the charge movement through DNA and studies of how such a charge movement can be accelerated [8] or manipulated are consequently active fields of research [9]. In this review we cover recent insights into the excess electron transfer capabilities of DNA, which were obtained with electron donor- and electron acceptor-modified oligonucleotides. In these oligonucleotides a reduced, deprotonated, and light-excited flavin functions as the electron donor. A cyclobutane pyrimidine dimer, which spontaneously cleaves upon single-electron reduction, is the electron acceptor. The special acceptor used for the studies allows one to translate the event of an incoming electron into a readily detectable strand break. A very similar sequence of events is used by the DNA repair enzyme DNA photolyase [10]. This enzyme uses a flavin-containing FAD cofactor to repair (split) cyclobutane pyrimidine dimer lesions in DNA. The lesions are formed in cellular DNA in response to sunlight irradiation. They are responsible for the development of sunlight-induced tumors and cell death [11]. A deeper understanding of the electron transfer-induced cleavage of a cyclobutane pyrimidine dimer by a flavin in DNA is consequently not only informative with respect to the question of how DNA mediates excess electron transfer but also may allow one to gain fundamental insight into one of nature's most important DNA repair pathways. In order to give the reader a comprehensive background on excess electron transfer in DNA, this review begins with a brief introduction into the DNA photolyase repair process. Subsequently we discuss electron transfer processes in DNA. This review covers only the transfer of excess electrons through DNA. This excess electron transfer process has to be distinguished from the traveling of a positive charge, of a hole,

Charge Transfer in DNA: From Mechanism to Application. Edited by H.-A. Wagenknecht

ISBN: 3-527-31085-1

through DNA. The first process is mediated by the pyrimidine bases, which are the easiest to reduce [12]. The second process, which is the subject of a number of excellent reviews [13–15], is guided by the purines due to their comparatively low oxidation potentials [16].

3.2
Excess Electron Transfer-driven DNA Repair by DNA Photolyases

Electron injection into DNA has been exploited by nature for millions of years. Presumably, since life exists on an earth exposed to sunlight, an enzyme called DNA photolyase [17] is used by many organisms to repair genotoxic UV-induced lesions [18, 19]. UV irradiation of cells induces the formation of a number of cyclobutane-type photoproducts by a photochemical [2π+2π] cycloaddition of adjacent pyrimidines in the double strand [20]. These lesions are enzymatically repaired (split) by electron injection from a light-excited, reduced, and deprotonated flavin coenzyme ($FADH^{-*}$) inside the protein. Upon single-electron reduction, the dimer (-T=T-) radical anion spontaneously cleaves back into the two monomers (repair: $\text{-T=T}^{\bullet -}\text{-} \rightarrow \text{-T T}^{\bullet -}\text{-}$) as shown in Scheme 3.1 [10].

This electron transfer process from a light-excited $FADH^{-*}$ to a cyclobutane pyrimidine dimer (T=T dimer) is a thermodynamically favorable process that has been modeled extensively with model compounds [21–24]. The reduction potential of the reduced and deprotonated $FADH^{-*}$ in its photoexcited state is believed to be around $E_{red^*} = -2.6$ V against NHE [25–27], which is negative enough to reduce all nucleobases (dG may be an exception) including the T=T dimer, as is evident

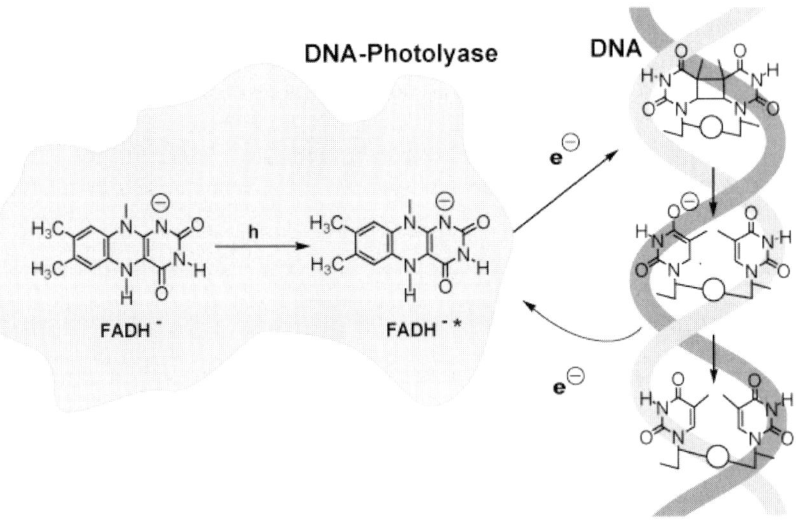

Scheme 3.1 Repair mechanism employed by DNA photolyases for the cleavage of a mutagenic cyclobutane-type lesion in DNA.

Table 3.1 Reduction potentials of some nucleobases.

	Reduction potentials (V)		
Base	**E(Red)** [a)]	**E(Red)** [b)]	**E(Red)** [c)]
dG	−2.76		
dA	−2.45		
dC	−2.23	−1.09	−2.1 (DMC)
dT	−2.14	−1.10	−2.1 (DMT)
U	−2.04	−1.05	−2.1 (DMU)
T=T			−2.2 (DMTD)

DMC, dimethylcytosine; DMT, dimethylthymine; DMU, dimethyluridine; DMTD, dimethylthymine dimer.

a) Polarographic potentials in DMF versus NHE [16].

b) Data from pulse radiolysis experiments in water at pH 8.5 against NHE [29].

c) Data from fluorescence-quenching experiments in acetonitrile against SCE [26, 27].

from the reduction potentials listed in Table 3.1. The reduction potentials show that thymine and the T=T dimer are the DNA bases that are most easily reduced. An electron injected into DNA will therefore trigger an electron transfer process within the duplex to either a thymine base or a cyclobutane thymine dimer lesion if present. In an aqueous environment, the pK_a values of the nucleobase radical anions also need to be considered, because proton transfer will have a strong impact on the reduction potentials. Table 3.2 lists the pK_a values of some nucleobase radicals [28]. The thymine radical anion has a rather neutral pK_a of about 7. Protonation of the cytosine radical anion (pK_a = 13), however, is strongly exothermic. An excess electron deposited on a cytosine may therefore trigger rapid protonation of the radical anion, which would trap the excess electron on the cytosine and disrupt any excess electron transfer through the duplex. The data therefore indicate that a reduced and deprotonated flavin, if light-excited, is able to inject an electron into a DNA double strand roughly regardless of the sequence context. This electron should be able to travel through the DNA strand. If it encounters a cyclobu-

Table 3.2 pK_a values of the reduced nucleobases [28, 30].

Equilibrium	**pK_a**
$T^{\bullet} \rightleftharpoons [T\text{-}H^+]^{\bullet -} + H^+$	6.9
$TH^+ \rightleftharpoons T + H^+$	−5
$A \rightleftharpoons [A\text{-}H^+]^- + H^+$	>14
$C^{\bullet} \rightleftharpoons [C\text{-}H^+]^{\bullet -} + H^+$	13
$CH^+ \rightleftharpoons C + H^+$	4.3
$G \rightleftharpoons [G\text{-}H^+]^- + H^+$	9.5

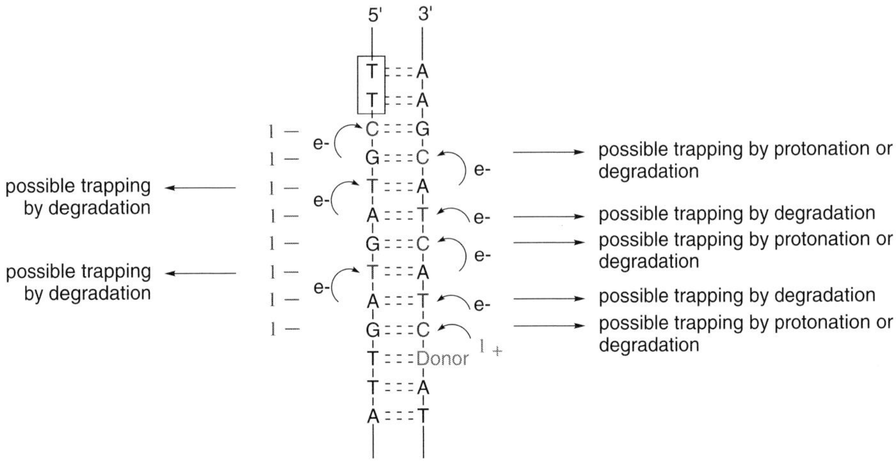

Fig. 3.1 Depiction of the possible sequence of events encountered during excess electron hopping transfer through DNA.

tane pyrimidine dimer, it should induce ring opening. The excess electron may also travel over C-G base pairs, but it may be trapped by cytosine radical anion protonation. Alternatively it can trigger a base decomposition process leading to the formation of reductive DNA lesions under consumption of the excess electron.

Figure 3.1 summarizes our current view about an excess electron transfer process through DNA. A donor injects an electron into DNA, which travels through the base stack using the pyrimidine bases as mediators due to the higher reduction potentials compared to the purine bases.

3.3 Excess Electron Transfer in DNA

3.3.1 Distance Dependence

Today our knowledge about excess electron transfer is still rudimentary due to the lack of a broad range of different model systems that would allow a deep comparative investigation of the process from different point of views. The available data were gained from direct radiolysis studies of DNA and DNA modified with intercalators [31], studies of pyrene-modified oligonucleotides [32] in combination with short-time spectroscopy [33], and studies of two defined donor-DNA-acceptor systems. The system employed by Rokita utilizes an aromatic amine as the electron donor and a bromouracil as the electron acceptor, which after electron reduction leads to a cascade of reactions ultimately yielding a strand break [34]. The first defined model system used to study excess electron transfer in DNA possessed a flavin electron donor and a thymine dimer acceptor, which cleaves spontaneously after single-electron reduction [35]. In order to investigate over which distances an excess electron may be able to travel in DNA, a series of DNA double strands were prepared that contained a special T=T lesion and a flavin donor specifically incorporated in DNA [36]. The dimer possesses an open backbone, inducing an easily detectable strand break upon single-electron reduction. The chemistry that was used to investigate the electron transfer event is depicted in Scheme 3.2. The DNA double strands **1a–1h** prepared for the study are shown in Figure 3.2. In these double strands, the distance between the flavin donor and the dimer accep-

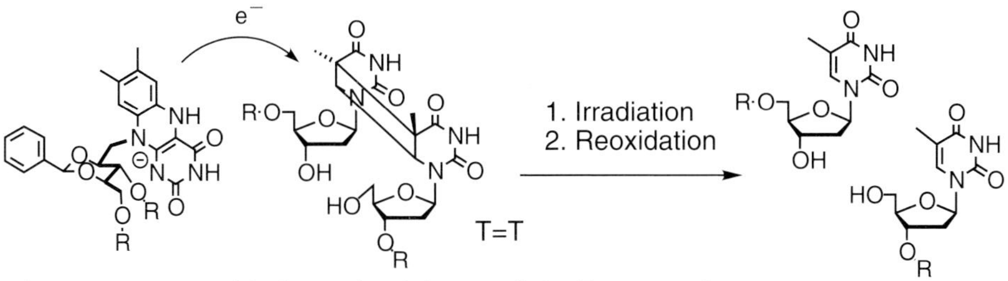

Scheme 3.2 Depiction of the flavin-induced cleavage of a backbone-opened T=T, which gives rise to a DNA strand break.

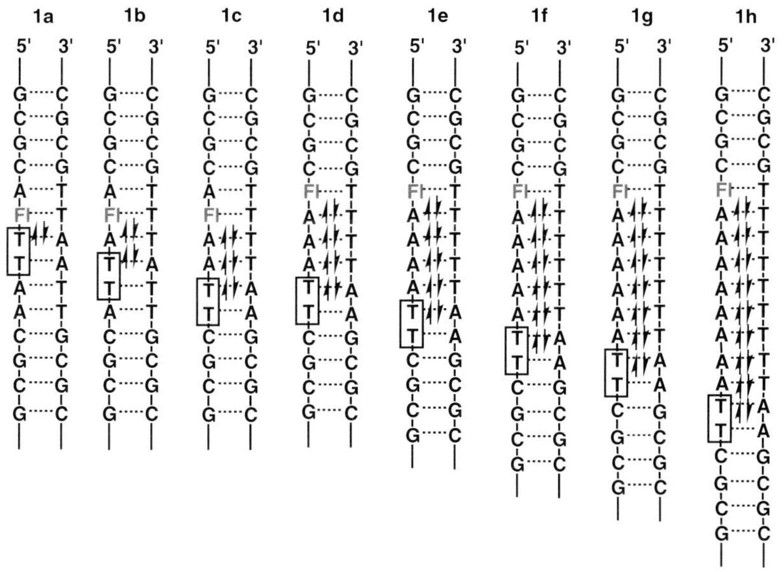

Δr (Å)	3.4	6.8	10.2	13.6	17	20.4	23.8	30.6
N	1	2	3	4	5	6	7	9
d (%/min)	3.7	2.5	1.0	2.2	1.9	1.0	0.8	0.13

Fig. 3.2 Depiction of the DNA double strands containing the flavin and the T=T dimer (TT) at systematically increased distances [36].

tor was systematically increased from 3.4 Å to about 30 Å using additional A-T base pairs between the redox partners. For the measurement, solutions (pH = 7.4) containing the DNA strands were irradiated (360 nm) under anaerobic conditions after reduction of the flavin with sodium dithionite [36]. HPLC analysis of samples removed from the assay solution after defined time intervals allowed us to quantify the amount of dimer splitting depending on the irradiation time for each of the DNA duplexes shown in Figure 3.2.

Since efficient repair (splitting) of the dimer was observed even in the double strands **1g** and **1h**, the experiments showed that an excess electron can travel over a distance of about 30 Å [36]. Furthermore, the rather small decrease of the repair efficiency with increasing distance indicated that the excess electron does not travel directly from the flavin to the dimer using a Marcus-type mechanism but rather hops over the intermediate A-T base pairs as a temporary charge carrier (see below). This result is in good agreement with data obtained from γ-radiolysis and EPR experiments [31, 37, 38].

In order to confirm the conclusions made with flavin- and thymine dimer-modified, double-stranded DNA, four DNA hairpins (**2a–2d**) were prepared in which the distance between the flavin donor and the dimer acceptor was again systematically increased (6.8 Å to 17 Å) (Figure 3.3). These hairpin structures contain the flavin molecule **3** as a cap [39]. They consequently possess only one helix-disturbing unit, the dimer, inside the DNA duplex region. This was considered to be an advantage because two disturbing units might have hindered the DNA in adopting a B-type double-helix conformation.

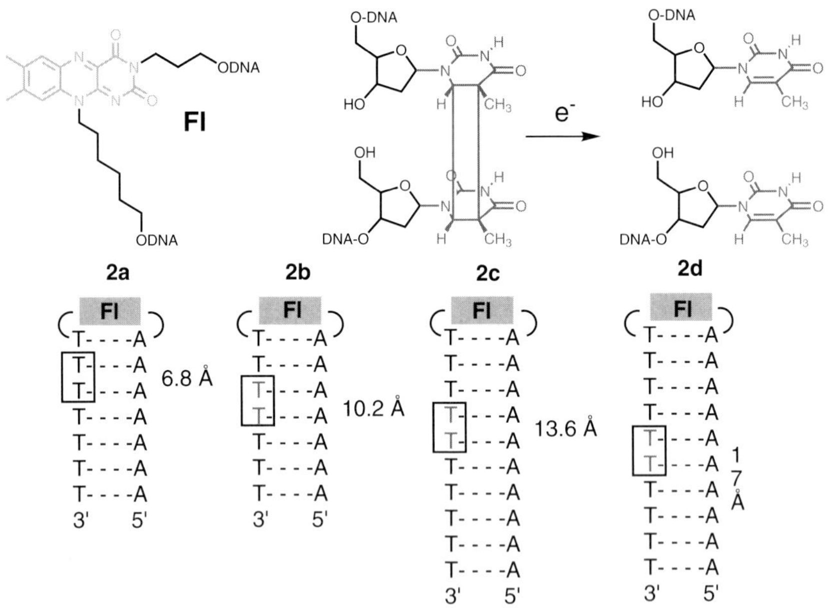

Fig. 3.3 Depiction of the DNA hairpins used to study the distance dependence of excess electron transfer [39].

Table 3.3 Melting points of DNA hairpins **2a–2d** and determined repair yield after 1 min of irradiation[a)].

Hairpin	2a	2b	2c	2d
Melting point (°C)	28	12	36	30
Repair yield (percentage per min) (5 °C)[b]	2.0	0.5	0.5	0.1
Repair yield (percentage per min) (0 °C)[b]	2.9	0.7	1.2	0.1

a) *c*DNA = 20 μM, 0.01 M Tris, pH 7.4, 150 mM NaCl.
b) Repair yield determined by ion-exchange chromatography: Nucleogel SAX 1000–8 (VA 50/4.6), pH 12. Linear gradient of 0.2 M NaCl to 1 M NaCl over 35 min.

The data in Table 3.3 show that dimer cleavage proceeds efficiently in hairpins **2a–2d**. Excess electron transfer decreases with increasing distance, from 2% repair per minute at 5 °C in **2a** to about 0.1% repair per minute in hairpin **2d**. If we consider that the distance between the two redox partners increases in these hairpins from 6.8 Å to about 17 Å, we can conclude that the distance dependence is again low and hence is not in agreement with a Marcus-type behavior. Overall, however, the distance dependence of the excess electron transfer is more pronounced than hole transfer, which is in good agreement with a recent short-time spectroscopic study and with data from Rokita [34, 40–42].

Marcus-type electron transfer is exponentially distance dependent, with $k_{ET} = A\exp(-\beta'\Delta r)$. For DNA (hole transfer), β' values between 0.7 $Å^{-1}$ and 1.2 $Å^{-1}$ were determined [13, 43]. In the Marcus model, one would expect a decrease of the repair yield by a factor of about eight with every additional base pair introduced between the dimer and the flavin. This would predict for hairpin **2d**, in comparison to **2a**, a repair yield of about 0.004% per minute, which is 1–2 orders of magnitude lower than observed. A plot of our yield data obtained with hairpins **2a–2d** (ln y) against the distance Δr (Figure 3.4) provided a very low β' value of about 0.3 $Å^{-1}$ at 5 °C and at 0 °C, showing that a direct Marcus-type electron transfer from the fla-

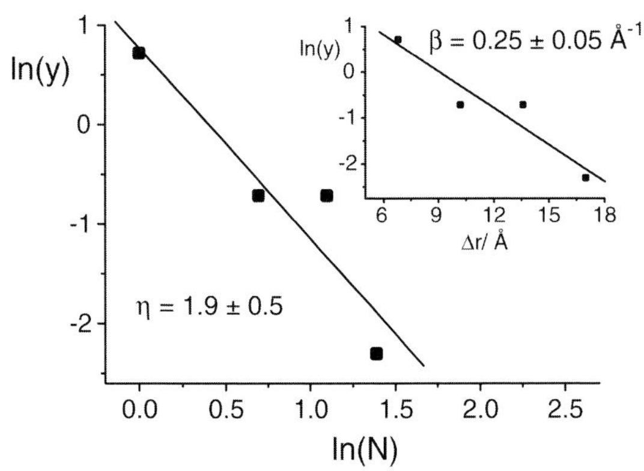

Fig. 3.4 Plots of the obtained data using the Marcus model [ln(y) against Δr] and the hopping model [ln(y) against ln(*N*)].

vin to the dimer is unlikely. In the hopping model, the electron is not directly transferred but uses intermediate charge carriers to hop from the donor to the acceptor. In the hopping scenario, the transfer efficiency is much less distance dependent. With $\ln(k_{ET}) = -\eta \ln(N)$, the transfer rate is proportional to the number of hopping steps (N) [43]. The proportionality factor η should be around two if the electron moves in a random walk-like process [43, 44]. If we assume that the electron uses the T bases in every A-T base pair as a stepping stone, a plot of the measured cleavage yields per minute $\ln(y)$ against $\ln(N)$ (Figure 3.4) provided an η value close to 2. This is in excellent agreement with the hopping model and with the data obtained with the DNA double strands. Here the mathematical analysis in the hopping mode also provided an η value of around 2. Analysis of the data assuming Marcus-like direct electron transfer again gave a too small value for β' of about 0.1 $Å^{-1}$ [36]. The β' value of about 0.3 $Å^{-1}$ from the hairpin studies is in excellent agreement with recent data from Rokita [34].

3.3.2
Directional Dependence

If the electrons move in a random walk-like hopping process, one expects that the excess electron travels along a DNA duplex in both directions (5′3′ versus 3′5′) with the same efficiency. This was probed with a series of PNA-DNA [45] double strands, which contain the flavin electron donor in the PNA strand and the dimer in the DNA counter-strand [46, 47]. All investigated duplexes are listed in Table 3.4.

In the DNA-PNA hybrids **3a–3d**, the electron travels from 5′ to 3′ over distances of about 3.4 Å (**3a**), 10.2 Å (**3b**), 17.0 Å (**3c**), and 23.8 Å (**3d**). In DNA-PNA hybrids **3e–3h**, the electron transfer has to hop in the opposite direction (3′ → 5′) over the same distances. The base sequence between the flavin donor **F** and the dimer acceptor **T̲T̲** was kept constant.

The measured cleavage yields obtained after irradiation of all hairpins for 1 min are listed in Table 3.5 together with the melting points of the duplexes. The yield data show that the dimer cleavage proceeds efficiently in all investigated DNA-PNA double strands **3a–3h** even over a distance of about 24 Å [47]. In agreement with the studies reported before, one observes that the dimer cleavage is not very distance dependent, even in this intermolecular case. In fact, the distance dependence of these DNA-PNA hybrids is, with a factor of only about 2, almost negligible. We also consistently observed that the DNA-PNA hybrids with the largest flavin–thymine dimer separation gave slightly more efficient dimer cleavage. We currently believe that the larger separation of the two unnatural bases (**F** and **T̲T̲**) allows the duplex to adopt, in between both disruptive elements, a more ordered duplex structure [46].

If we compare the cleavage data obtained within the PNA-DNA series **3a–3d** (electron transfer in the 5′ → 3′ direction) with the series **3e–3h** (electron transfer in the 3′5′ direction), we observe no large repair yield difference, showing that the repair of a thymine dimer, even over rather large distances of 24 Å by a reduced and deprotonated flavin, is independent of the electron transfer direction. This

Table 3.4 DNA-PNA hybrid double strands **3a–3h** used to study the directional and distance dependence of the excess electron transfer process. PNA bases are shown in italics.

Number	Name	Sequence
3a	DNA	5′-GCA-AAA -AAA- ATT-CGC-3′
	PNA	*KK-CGT-TTT -TTT -* **F***AA -GCG-KK-*NH_2
3b	DNA	5′-GCA-AAA -AAA- ATT-CGC-3′
	PNA	*KK-CGT-TTT -T***F***T - TAA -GCG-KK-*NH_2
3c	DNA	5′-GCA-AAA -AAA- ATT-CGC-3′
	PNA	*KK-CGT-TT***F** *-TTT - TAA -GCG-KK-*NH_2
3d	DNA	5′-GCA-AAA -AAA- ATT-CGC-3′
	PNA	*KK-CGT-***F***TT -TTT - TAA -GCG-KK-*NH_2
3e	DNA	5′-CGC-TTA-AAA- AAA -ACG-3′
	PNA	*KK-GCG-AA***F** *-TTT - TTT -TGC-KK-*NH_2
3f	DNA	5′-CGC-TTA-AAA- AAA -ACG-3′
	PNA	*KK-GCG-AAT -T***F***T - TTT -TGC-KK-*NH_2
3g	DNA	5′-CGC-TTA-AAA- AAA -ACG-3′
	PNA	*KK-GCG-AAT -TTT -* **F***TT -TGC-KK-*NH_2
3h	DNA	5′-CGC-TTA-AAA- AAA -ACG-3′
	PNA	*KK-GCG-AAT -TTT - TT***F** *-TGC-KK-*NH_2

K = lysines needed for solubility reasons.

Table 3.5 List of the dimer cleavage yields after irradiation of the PNA-DNA double strands **3a–3h** for 1 min[a].

Number	Melting point (°C)[b]	Distance (Å)	Direction	Yield (%)
3a	82	3.4	5′→3′	1.1±0.3
3b	67	10.2	5′→3′	1.2±0.3
3c	69	17.0	5′→3′	1.2±0.3
3d	70	23.8	5′→3′	1.3±0.3
3e	77	3.4	3′→5′	1.5±0.3
3f	–[c]	10.2	3′→5′	0.8±0.2
3g	68	17.0	3′→5′	1.8±0.4
3h	69	23.8	3′→5′	2.3±0.5

a) 1000 W Hg(Xe)-lamp, 10 °C, 360 nm cutoff filter.
b) 3 μM DNA, 3 μM PNA, 10 mM NaCl, 10 mM H_3PO_4-Na_2HPO_4 (pH 7.0).
c) Not measured.

result nicely supports that excess electron transfer in duplexes is indeed a random walk process in which the electrons hop along the DNA duplex without showing any directional preference. This result is currently not in agreement with data from the Rokita group, who observed faster excess electron transfer in the 3′ → 5′ direction [34]. Clearly, further studies are needed to resolve this controversy.

3.3.3
Sequence Dependence

The question of sequence dependence and in particular whether cytosines would interfere with the excess electron transfer process was investigated with another set of DNA hairpins depicted in Figure 3.5 [48]. These DNA hairpins (**4a–4e**) (Scheme 3.1) possess a different base sequence between the flavin donor, functioning as a cap, and the dimer acceptor. Hairpins **4a** and **4d** contain a homo-A-T stretch between the flavin donor and the dimer acceptor. In the hairpins **4b** and **4c**, one of the A-T base pairs is replaced by a G-C base pair at different positions. Finally, in hairpin **4e**, the flavin donor and the dimer acceptor are separated by a homo-G-C stretch. The distance between the donor and the acceptor in hairpins **4a–4c** is about 13.6 Å. This distance is increased in hairpins **4d** and **4e** to about 17.0 Å. This ensures that the electron transfer proceeds by charge hopping, where the intermediate base pairs function as charge carriers [43, 44].

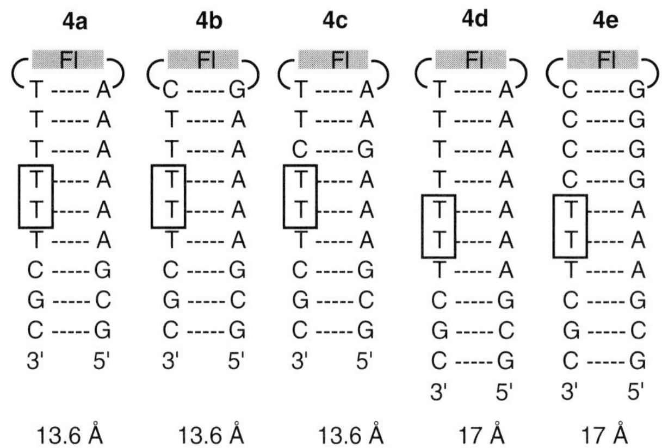

Fig. 3.5 Depiction of DNA hairpins **4a–4e** used to study the sequence dependence of the excess electron transfer. Fl = flavin, TT = cyclobutanethymidine dimer

Cleavage of the dimer by excess electron transfer proceeded efficiently in all these hairpins. In agreement with earlier studies, dimer cleavage is only about 30% slower in hairpins **4d** and **4e**, again showing the weak distance dependence of the repair reaction.

The data depicted in Table 3.6 show for hairpins **4a–4c** (flavin dimer distance = 13.6 Å) that the sequence between the flavin and the dimer does not influence the

Table 3.6 Melting points and cleavage yields determined for hairpins **4a–4e**.[a)]

	Hairpin				
	4a	**4b**	**4c**	**4d**	**4e**
Melting point (°C)	45	48	46	40	50
Yield (percentage per min)[b)]	**4.2**	**4.3**	**3.9**	**2.5**	**3.0**

a) *c*DNA = 3 μM, 0.01 M Tris, pH = 7.4, 0.15 M NaCl. The distance between the flavin and the dimer was estimated assuming ideal B-conformation of the hairpin stems.
b) ±20% as determined from three independent measurements. The data were determined from the first 5 min of irradiation after linear approximation of the data.

dimer-splitting yield. All three hairpins gave very similar repair yields during the experiment. Most interesting is a direct comparison of the electron transfer through the homo-A-T sequence in **4d** and transfer through the homo-G-C sequence in **4e**. One clearly observes no strong sequence effect. In fact, repair through a homo-A-T stretch gives reproducibly slightly lower splitting yields. The reason for this effect again could be a more flexible hairpin structure, which would be in agreement with the lower melting point measured for hairpin **4d**. The surprising result reported in this study is that the reductive cleavage of a thymine dimer in DNA hairpins over a distance of 17 Å is not sequence dependent [48]! This result is also not in agreement with measurements by Rokita et al., who reported less efficient transport of an excess electron through a G-C base pair [34, 42].

3.4
The Catalytic Electron? Or, Can One Electron Repair More Than One Dimer Lesion?

After cleavage of the thymine dimer using an excess electron, one neutral thymine and one thymine radical anion are formed. In principle, this extra electron either can be trapped or it can continue its journey through the DNA duplex and cleave another DNA lesion at another position. In order to investigate the fate of the injected electron, investigations were performed with the electron donor developed by Giese and coworkers [49]. In contrast to the flavin electron injector, the Giese electron injector **5** transfers upon irradiation only a single extra electron into the DNA double strand. The basis for the new injector is the less negative redox potential of thymine [16, 29, 50] compared to a dialkyl ketone radical [51]. Thus, a ketyl radical anion produced by a Norrish type 1 cleavage in close proximity to a thymine was used to prepare a single thymine radical anion inside a DNA duplex (Scheme 3.3).

For the experiment, the double strand **8** (Scheme 3.4) containing two thymine dimers separated by a single A-T base pair was prepared. Upon irradiation of

4
Dynamics of Photoinitiated Hole and Electron Injection in Duplex DNA

Frederick D. Lewis and Michael R. Wasielewski

4.1
Introduction

The structure, biological function, and physical properties of duplex DNA have long been the subject of fascination, speculation, and investigation for biological and physical scientists. From a structural perspective, the unique features of DNA are its hydrophilic, helical sugar-phosphate backbone and its hydrophobic, hydrogen-bonded base pairs, which constitute the core of DNA [1, 2]. The base pairs form an extended, one-dimensional π-stacked array with an average stacking distance of 3.4 Å. Weak excitonic interactions between base pairs are responsible for the characteristic UV absorption and circular dichroism spectra of duplex DNA [3].

The possibility that the π-stacked base pairs of DNA might function as a one-dimensional conductor or molecular wire was first advanced by Eley and Spivey [4] shortly after Watson and Crick deduced the double-helical structure of DNA [5]. Pulsed radiolysis studies of DNA at low temperatures established that oxidation of DNA results in the formation of "holes" localized on guanine, the most readily oxidized of the nucleobases, and that reduction of DNA results in electron attachment to thymine or cytosine, the most readily reduced of the nucleobases [6]. These studies also indicated that both holes and electrons can migrate over short distances prior to being trapped by the appropriate base.

Interest in the wire-like behavior of DNA was kindled in the early 1990s by a series of papers by Barton and Turro suggesting that ultrafast, photoinitiated electron transfer between intercalated electron donors and acceptors could occur over long distances in DNA [7–9]. These reports stimulated experimental studies of photoinitiated electron transfer processes in DNA as well as their theoretical interpretation. DNA electron transfer has since been the subject of numerous commentaries and reviews and no small measure of controversy [10–18].

In 1997 we reported our initial studies of the distance-dependent dynamics of hole injection and charge recombination in synthetic DNA hairpins [19]. In these experiments, the singlet state of a stilbenedicarboxamide hairpin linker serves as the hole donor and a guanine located in the base-paired hairpin stem as the hole

Charge Transfer in DNA: From Mechanism to Application. Edited by H.-A. Wagenknecht

ISBN: 3-527-31085-1

acceptor. By varying the linker acceptor and the duplex base sequence, it has been possible to investigate the dynamics of hole injection as a function of distance [20], driving force [21], and donor-bridge-acceptor energetics [22]. The dynamics and equilibria for single-step hole transport processes have been studied in hairpins possessing both primary and secondary hole acceptors [23]. The dynamics of electron injection and charge recombination has also been investigated using an electron donor linker and nucleobases as the electron acceptors [24]. We present here a review of our contributions to the study of the dynamics of electron transfer processes in DNA and some comparisons to relevant experimental and theoretical studies from other laboratories.

4.2
DNA Hairpin Synthesis, Structure, and Energetics

4.2.1
Hairpin Synthesis and Structure

The preparation of synthetic DNA conjugates possessing a stilbenedicarboxamide (**Sa**) linker connecting short, complementary oligonucleotides employs the method initially reported by Letsinger and Wu (Scheme 4.1) [25]. Briefly stated, a diol derivative of the linker chromophore is first monoprotected (as its (dimethoxytriphenylmethyl)ether [DMT] derivative) and then activated (as the phosphoramidite) prior to incorporation into a bis(oligonucleotide) conjugate using standard phosphoramidite chemistry. This general procedure has been used to prepare hairpin structures possessing a variety of linker chromophores, selected examples of which are shown in Scheme 4.2.

Conjugates possessing as few as three or four A-T base pairs or one or two G-C base pairs form stable hairpin structures, whereas analogous hairpins possessing

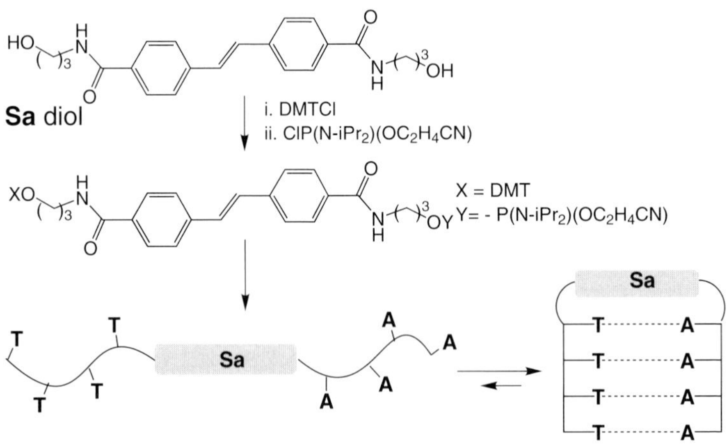

Scheme 4.1 Synthesis of **Sa**-linked hairpins.

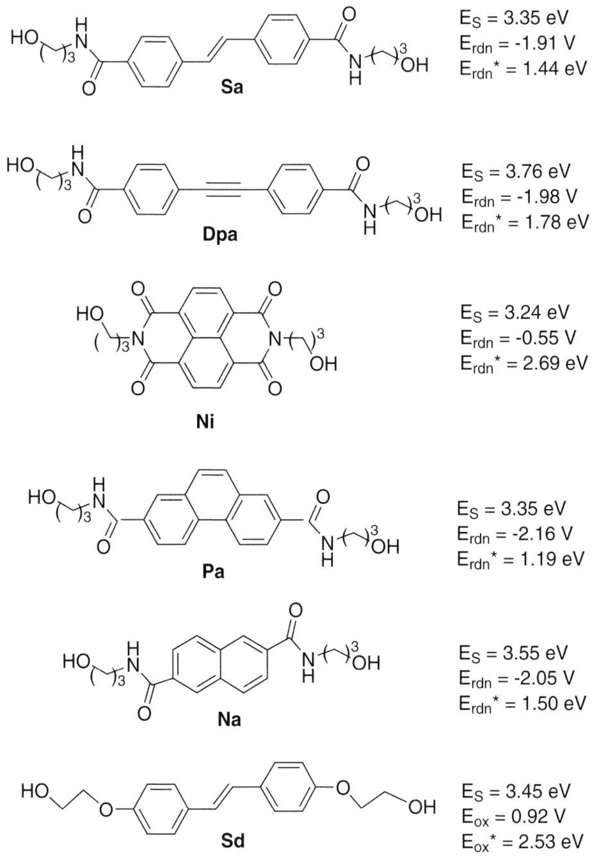

Scheme 4.2 Structures, singlet energies, and redox potentials of hairpin linkers as their diol precursors.

polynucleotide loop regions are unstable at room temperature [26]. The enhanced thermal stability of the synthetic hairpins is a consequence of the entropic advantage of a compact, rod-like aromatic core joined to the sugar-phosphate backbone with short, flexible connectors. Hydrophobic association of the stilbene chromophore and adjacent base pair also contributes to hairpin stability. In collaboration with Egli and coworkers, we have determined the crystal structures of a hairpin possessing a stilbenediether (**Sd**) linker and six base pairs with both Sr^{2+} and Mg^{2+} counterions [27, 28]. The structure of one of four hairpins from the unit cell of the low-resolution Sr^{2+} structure is shown in Figure 4.1. The **Sd** chromophore is parallel to the adjacent G-C base pair, and the duplex domain adopts a classic B-form structure.

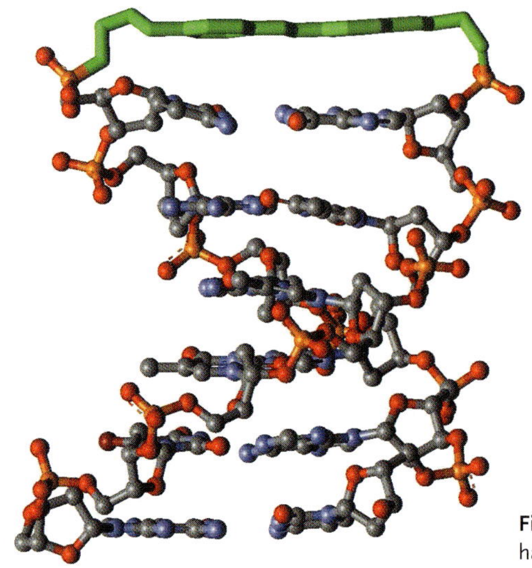

Fig. 4.1 Crystal structure of an **Sd**-linked hairpin with stilbene linker at top.

4.2.2
Electron Transfer Energetics

The free energy of a photoinduced charge separation process in which the electronically excited hairpin linker serves as either an electron acceptor (Ac) or a donor (D) and the nucleobase (N) serves as an electron donor (hole acceptor) or an electron acceptor, respectively, can be estimated from the redox potentials of the donor and acceptor using the Rehm-Weller relationship (Eqs. (1) and (2)):

$$\Delta G_{cs} = E^0(N^{+\bullet}/N) - E^0(Ac/Ac^{-\bullet}) - E^*(Ac) + \Delta G^0(\varepsilon) \tag{1}$$

$$\Delta G_{cs} = E^0(D^{+\bullet}/D) - E^0(N/N^{-\bullet}) - E^*(D) + \Delta G^0(\varepsilon) \tag{2}$$

$$\Delta G_{cr} = E^0(D^{+\bullet}/D) - E^0(N/N^{-\bullet}) \text{ [or } E^0(N^{+\bullet}/N) - E^0(Ac/Ac^{-\bullet})], \tag{3}$$

where, E^* is the singlet energy of the excited state and $\Delta G^0(\varepsilon)$ is a solvent-dependent correction term that is small (ca. -0.1 V) in water and polar organic solvents [29]. The ion pair formed in the charge separation process can return to the ground state by means of return electron transfer (charge recombination). The free energy for this process is approximately equal to the sum of the redox potentials (Eq. (3)).

Values of E^* for fluorescent molecules can readily be assigned from the wavelength of the high-energy onset of the fluorescence emission spectrum. Ideally, the redox potentials of the electron donor and acceptor would be determined directly in duplex DNA. However, this is not possible experimentally. We have consistently employed linker and nucleotide oxidation and reduction potentials determined in polar, aprotic solvents such as acetonitrile or dimethyl sulfoxide [30], assuming that these solvents provide an environment more similar to that of the base-paired core of DNA than would water. The linker singlet energies, the redox

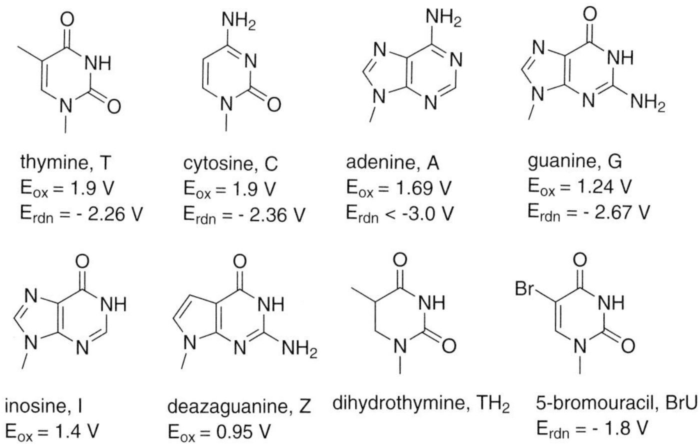

Scheme 4.3 Structures and redox potentials of the nucleobases.

potentials, and the sum of these values (the excited-state redox potentials) are reported in Scheme 4.2 and the structures and potentials for the bases employed in our studies are shown in Scheme 4.3 [31].

The singlet energies and redox potentials can be utilized to construct state energy diagrams for electron transfer processes. For example, hole transfer from singlet **Sa** to G is exergonic, whereas hole transfer from **Sa** to A (or T or C) is endergonic (Scheme 4.4 a), in accord with the selective photooxidation of G by **Sa** (Section 4.3.1) [19]. Electron transfer from singlet **Sd** to either T or C is exergonic, whereas electron transfer to G is endergonic (Figure 4.4 b), in accord with the ability of a GG base pair to inhibit electron transfer from Sd* to T (Section 4.5.2) [24]. In view of the assumptions made in the selection of nucleobase potentials, values of ΔG_{cs} and ΔG_{cr} obtained using Weller's equation should be regarded as approximations.

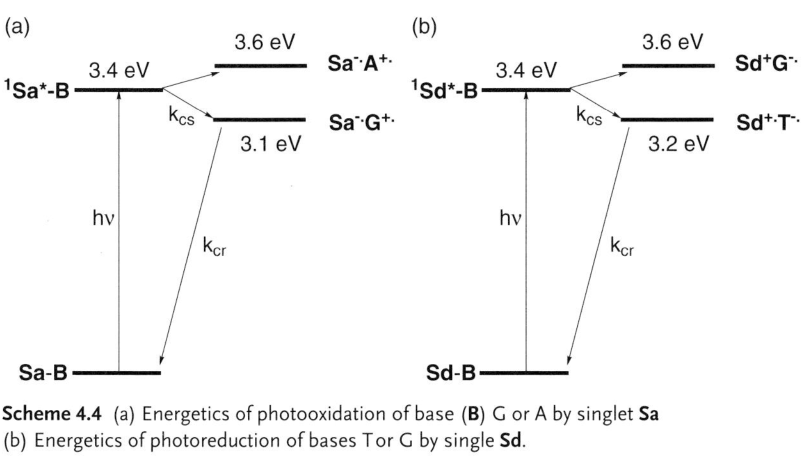

Scheme 4.4 (a) Energetics of photooxidation of base (**B**) G or A by singlet **Sa**
(b) Energetics of photoreduction of bases T or G by single **Sd**.

4.2.3
Electron Transfer Dynamics

The absorption and fluorescence spectra of the **Sa** linker and hairpin **1**, which has a base pair domain consisting of six A-T base pairs, are shown in Figure 4.2 [20]. Since the base pair domain is non-fluorescent and does not absorb at wavelengths longer than 300 nm, it is possible to selectively excite the **Sa** chromophore and to monitor its fluorescence or transient absorption without interference from the base pairs. Introduction of a single G-C base pair near the **Sa** linker results in reduction of the **Sa** fluorescence intensity and lifetime. Fluorescence quenching is attributed to an electron transfer mechanism, in accord with the energetics of hole injection on guanine (Scheme 4.4 a).

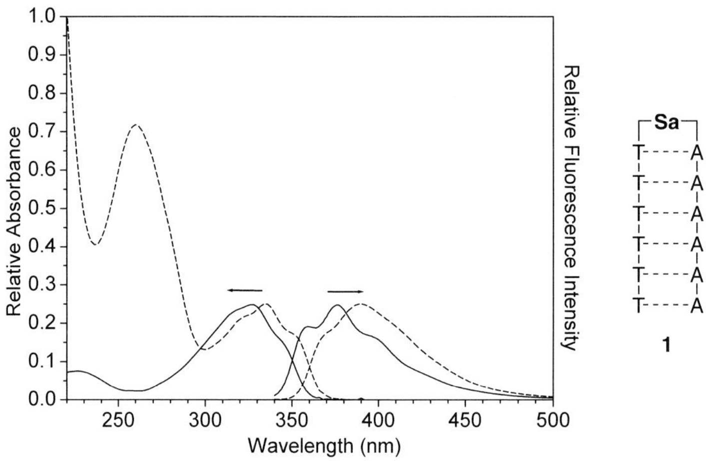

Fig. 4.2 Absorption and fluorescence spectra of the stilbene diol linker **Sa** (solid lines) in methanol solution (7.7×10^{-6} M) and the **Sa**-linked hairpin **1** (dashed lines) in aqueous solution (4.9×10^{-6} M, 0.1 M NaCl, 10 mM sodium phosphate, pH 7.2).

The dynamics of charge separation and charge recombination have been investigated by means of femtosecond and nanosecond time-resolved transient absorption spectroscopy [20]. The time-resolved spectra of the **Sd**-linked hairpin **2**, which possesses an adjacent G-C base pair, are shown in Figure 4.3 [24]. The 1-ps spectrum grows in during the laser pulse and is assigned to the singlet **Sd***. The 575-nm band decays with a time constant of ca. 3 ps accompanied by the formation of a narrower band at 535 nm assigned to the cation radical $\mathbf{Sd}^{+\cdot}$. The rate constant for charge separation (k_{cs}) can be calculated from the singlet decay times in the presence and absence of quenching (τ_s and τ_o, respectively ($k_{cs} = \tau_s^{-1} - \tau_o^{-1}$). The 535-nm band decays with a time constant of 40 ps, assigned to charge recombination of the radical ion pair. The rate constant for charge recombination k_{cr} is obtained directly from the radical ion pair decay time ($k_{cs} = \tau_{rip}^{-1}$). For some sys-

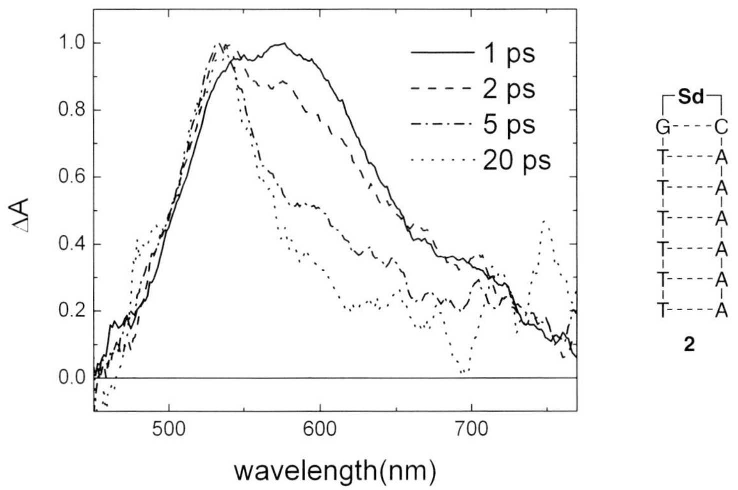

2

Fig. 4.3 Normalized transient absorption spectra for an **Sd**-linked hairpin obtained at several delay times following excitation at 340 nm with a 130-fs laser pulse.

tems, charge recombination occurs on the nano- to microsecond time scales, necessitating the use of a different laser apparatus. In cases where fluorescence lifetimes are also available, they are in reasonable agreement with the decay components assigned to the charge separation process.

4.3 Hole Injection

4.3.1 Distance Dependence

Our initial studies of the dynamics of charge separation (hole injection) and charge recombination employed two "families" of **Sa**-linked hairpins possessing a poly(A-T) base pair domain with a single G-C base pair, with G located either in the poly(T) strand (Figure 4.4, **3–7**) or in the poly(A) strand (*vide infra*) [19]. Since singlet **Sa** cannot oxidize A (Scheme 4.4 a), the A-T base pair domain separating the linker and G serves as a "bridge" between the donor and acceptor. Plots of log (k_{cs}) and log (k_{cr}) vs. the distance between the **Sa** linker and G are shown in Figure 4.5. Both k_{cs} and k_{cr} are seen to decrease as the distance R between **Sa** and G-C increases. In accord with a superexchange mechanism for photoinduced bridge-mediated electron transfer (Figure 4.4), the distance dependence can be described by:

$$k_{et} = k_o \exp(-\beta R) \quad (4)$$

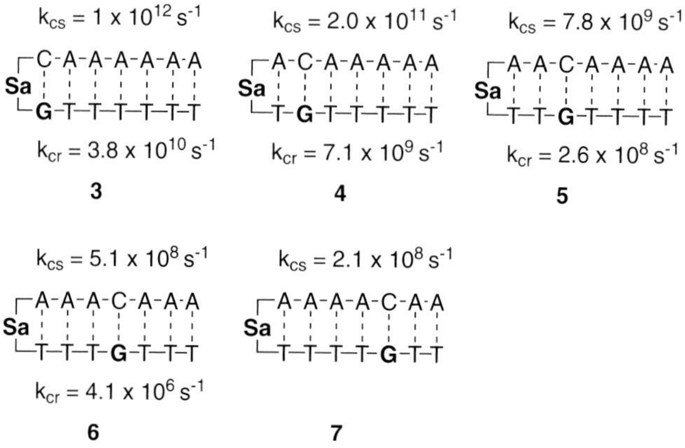

Fig. 4.4 Rate constants for charge separation and charge recombination for a family of **Sa**-linked hairpins.

where β is dependent upon the nature of the bridge and its coupling with the donor and acceptor. The data in Figure 4.5 provide values of $\beta_{cs} = 0.66$ Å^{-1} and 0.71 Å^{-1} and $\beta_{cr} = 0.90$ Å^{-1} and 0.94 Å^{-1} for the two hairpin families.

A value of $\beta \sim 0.7$ Å^{-1} indicates that duplex DNA can function as a more effective medium for electron transfer than a protein ($\beta \sim 1.0$–1.4 Å^{-1}) but that it is less effective than fully conjugated bridges that display wire-like distance depen-

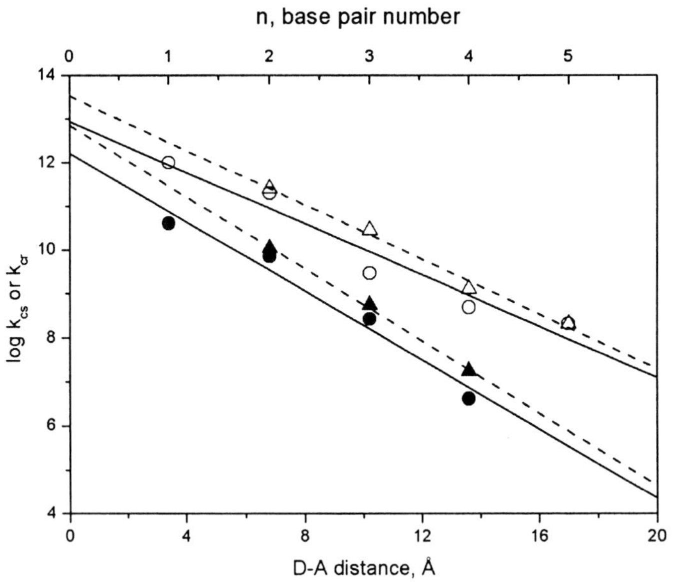

Fig. 4.5 Distance dependence of the rate constants for charge separation (open symbols) and charge recombination (filled symbols) for **Sa**-linked hairpin families in which guanine is in either the polyT arm (circles, solid lines) or the polyA arm (triangles, dashed lines).

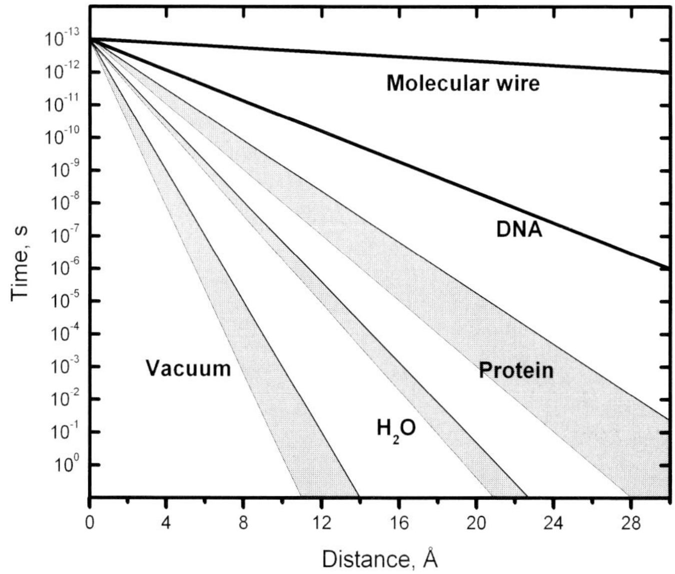

Fig. 4.6 Tunneling dynamics of bridge-mediated electron transfer (after [32]).

dence (Figure 4.6) [32, 33]. Several years after our initial report, Zewail and Barton [34] reported a similar value of β extracted from their analysis of the complex femtosecond singlet decay dynamics of duplex systems possessing an aminopurine singlet electron acceptor and guanine electron donor. Substantially greater distance dependence has been reported by Fukui and Tanaka and by Michel-Beyerle and coworkers for systems with an intercalated acridinium electron acceptor and guanine or deazaguanine as electron donors, indicative of values of $\beta > 1.5$ Å^{-1} [35, 36]. Values of $\beta > 1.0$ Å^{-1} have also been reported by Barbara and by Harriman from their analysis of DNA modified with donor and acceptor intercalators [37–39]. The large variation in the value of β for DNA-bridged systems is addressed in Section 4.3.3.

4.3.2
Electron Donor-Acceptor End-capped Hairpins

A common element in our studies and those of Zewail and Barton [34] and of Michel-Beyerle [36] is the use of G or Z bases as the hole acceptor. The weak transient absorption of G^+ prevents simultaneous monitoring of both components of the radical ion pair in these studies. This limitation led us to explore the use of the end-capped hairpins **8–10** (Figure 4.7) to investigate the distance dependence of electron transfer between a donor and acceptor stilbenes separated by a variable number of A-T base pairs [40]. These end-capped hairpins exhibit exceptional thermodynamic stability attributed to hydrophobic association of the **Sd** group with the terminal base pair in the duplex domain. Analysis of their circular dichroism

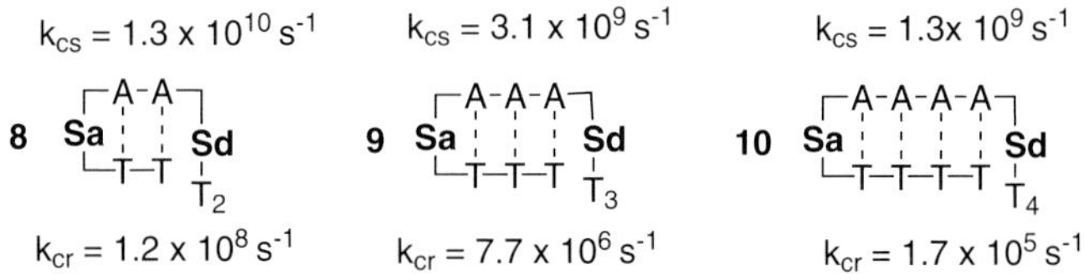

Fig. 4.7 Dynamics of charge separation and charge recombination for some D-A end-capped hairpins.

spectra indicates that the end-capped hairpins adopt B-DNA structures with both chromophores approximately perpendicular to the duplex axis.

The transient absorption spectra of the end-capped hairpins are consistent with the initial population of **Sa*** upon excitation at 360 nm, where **Sa** absorbs more strongly than **Sd**. The decay of the transient spectrum of **Sa*** is accompanied by the appearance of two narrower bands attributed to $\mathbf{Sa}^{-\bullet}$ and $\mathbf{Sd}^{+\bullet}$. Rate constants for charge separation and charge recombination for **8–10** obtained from the growth and decay of the $\mathbf{Sd}^{+\bullet}$ transient are summarized in Figure 4.7 [40]. Values roughly parallel those for **Sa**-linked hairpins with G donors (Figure 4.4). The calculated value of β_{cs} for the end-capped hairpins is somewhat smaller than for hairpins with G donors; however, the values of β_{cr} are similar.

4.3.3
Variation of the Tunneling Energy Gap

We have investigated the distance dependence of k_{cs} for hairpin families possessing phenanthrene dicarboxamide acceptor linkers (**Pa**, Scheme 4.2) and guanine or deazaguanine electron donors (Scheme 4.3) [22]. These systems also display stronger distance dependence than the **Sa**-linked hairpins. The value of β is found to increase as the energy gap, ΔE_{tun}, between the initial state and bridge-localized orbitals increases. According to the McConnell model, the distance dependence for bridge-mediated electron transfer is dependent upon the tunneling energy gap, which is smaller than ΔE_{tun} by an amount equal to half the energy bandwidth of the bridge eigenstates.

ΔE_{tun} is closely related to the "injection free energy", ΔG_{inj}, which is defined as the *minimum* free energy difference between the state with the hole localized on the bridge (A^-B^+D) and the initial or the final state (A^*BD or A^-BD^+). ΔG_{inj} is an

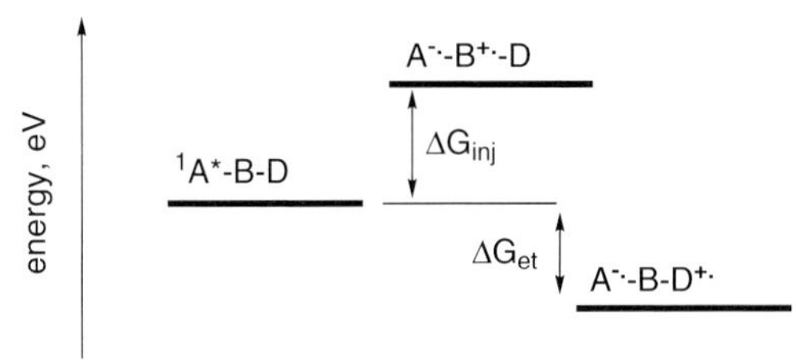

Scheme 4.5 Energetics of electron injection and electron transfer.

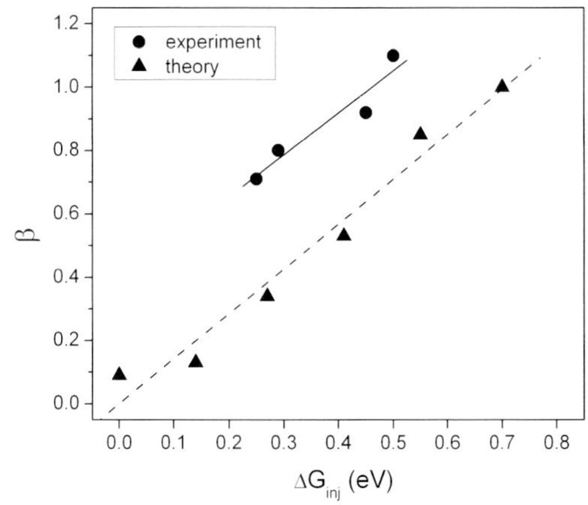

Fig. 4.8 Experimental and theoretical dependence of β on injection energy.

ensemble property and can be estimated from electrochemical data and excitation energies (Scheme 4.5). Both ΔE_{tun} and ΔG_{inj} increase as the bridge states are moved further in energy from the donor. A plot of experimental values of β for several hairpin systems vs. ΔG_{inj} is shown in Figure 4.8, along with the calculated values of Grozema et al. [41] for hole transport via a poly (A-T) bridge. The slopes of the experimental and theoretical plots are in good agreement.

In comparison to electron transfer via protein bridges, the variation of β-values for DNA bridge-mediated electron transfer is much larger [22]. This is a consequence of the much smaller tunneling energy gaps (or injection energies) of the π-orbitals of the aromatic bases, which constitute the bridging elements for DNA, vs. the σ-orbitals, which constitute the bridging elements for proteins. The rapid falloff in k_{cs} with distance for systems having an acridinium acceptor plausibly reflects strongly distance-dependent solvent reorganization energies for this ionic acceptor [35, 36].

4.3.4
Driving Force Dependence

The free energies for charge separation and charge recombination (Eqs. 1 and 2) can be varied by changing either the linker (Scheme 4.2) or the nucleobase (Scheme 4.3) [21]. When the nucleobase hole acceptor is adjacent to the linker, this can be accomplished for many combinations of linker and base. However, when the base is separated from the linker by one or more A-T base pairs, linkers such as **Sa** and **Pa** that do not oxidize A but oxidize G or Z must be employed. Plots of k_{cs} and k_{cr} vs. ΔG_{et} (Eqs. 1–3) are shown in Figure 4.9 along with global fits to the data obtained using the Marcus-Levich-Jortner (MLJ) theory. This analysis provides values of the solvent reorganization energy that are essentially independent of the donor-acceptor separation.

The results of our investigations of the distance and driving force dependence of k_{cs} have been analyzed by others using a variety of theoretical models. Tavernier

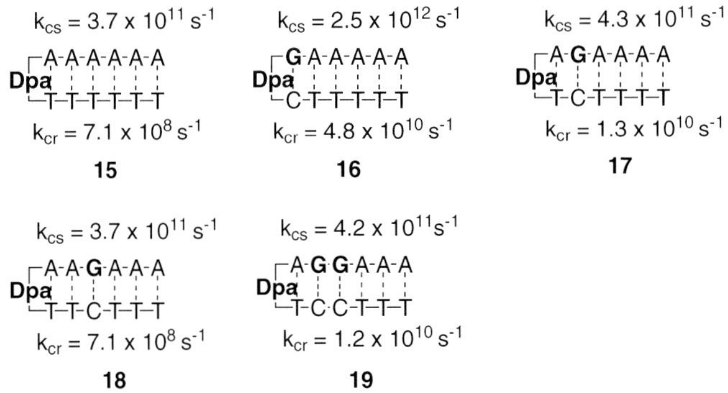

Fig. 4.11 Dynamics of charge separation and charge recombination for **Dpa**-linked hairpins with several base sequences.

ally formed contact radical ion pair fails to yield a base-separated ion pair is attributed to the distance dependence of the Coulomb attraction energy, as defined by:

$$\Delta G_{\mathrm{cm}} = \frac{e_0^2}{\varepsilon}\left(\frac{1}{r_{\mathrm{DA1}}} - \frac{1}{r_{\mathrm{DA2}}}\right) \tag{5}$$

where e_0 is the atomic unit of charge, ε is the solvent dielectric constant, and r_{DA1} and r_{DA2} are the distances between the contact and base-separated radical ion pairs, respectively. The values of r_{DA1} and r_{DA2} are estimated to be 3.4 Å and 6.8 Å, respectively, based on the average π-stacking distance in B-DNA. An upper bound for the value of $\Delta G_{\mathrm{cm}} \sim 0.4$ eV can be estimated from the difference between the oxidation potentials of G and A (Scheme 4.3) or from the difference in their adiabatic ionization potentials. The failure to observe hole migration from adenine to guanine suggests that the effective dielectric constant for the **DPA**-$A^{+\bullet}$ contact radical ion pair is <7. An important consequence of charge localization in the singlet contact radical ion pair is that charge recombination is both rapid and efficient.

4.4 Hole Transport

4.4.1 Overcoming Charge Recombination

In the preceding section we saw that hole transport cannot compete effectively with charge recombination in a contact radical ion pair, even when the hole transport process is expected to be exothermic. In order to overcome the Coulomb attraction in a contact radical ion pair, we investigated the possible occurrence of hole transport from a primary G donor to a secondary GG donor in hairpin systems possessing a variable number of A-T base pairs between the **Sa** linker and

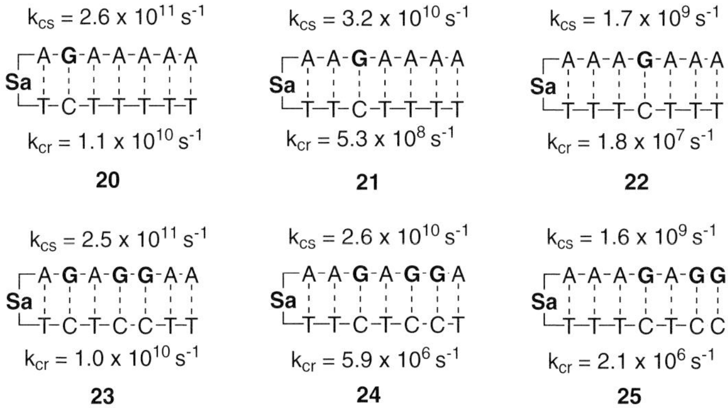

Fig. 4.12 Rate constants for charge separation and charge recombination for some **Sa**-linked hairpins. Values of k_{cr} for **20** and **21** are obtained from the longest-lived decay component ($k_{cr} = \tau_{rip}^{-1}$).

the primary donor. The results of our initial experiments provide the rate constants for charge separation and charge recombination for hairpins **20–25**, shown in Figure 4.12 [51]. As expected, the values of k_{cs} depend upon the distance between the primary donor and linker but are independent of the presence of the secondary donor. The values of k_{cr} are also independent of the presence of the secondary donor for the **Sa**-AGAGG base sequence in **23**. However, the presence of a secondary donor results in a significantly long-lived decay component for hairpins **24** and **25**. These results indicate that hole transport can compete with charge recombination when the later process is sufficiently slow ($<10^9$ s^{-1}).

4.4.2 Hole Transport Dynamics

The results of our experiments with secondary hole acceptors proved to be compatible with a kinetic model for decay of **Sa**$^{-\bullet}$ in which reversible hole transport competes with charge recombination in the primary radical ion pair (Scheme 4.6) [23, 48, 51–53]. Kinetic analysis provides values for the forward and return hole transport rates, k_t and k_{-t}. These values can be used to calculate the equilibrium constant and free energy for hole transport. Data for hairpins **24–33** are summarized in Table 4.1, in which the arm of the hairpin containing the primary donor is specified. Cross-strand hole transport is presumed to occur in hairpins **29** and **30**, which have a CC step at the end of the hairpin arm possessing the primary G donor. We note that reanalysis of the transient decay curves for sequence **28** provides values of k_t and k_{-t} that are an order of magnitude larger than previously reported. Other values remain essentially unchanged.

The values of k_t in Table 4.1 provide some interesting insights into the nature of the hole transport process. First, the rate constants for hole transport in GAGGA sequences **24** and **26** are slow compared to the rates of hole injection in the **Sa**-AG

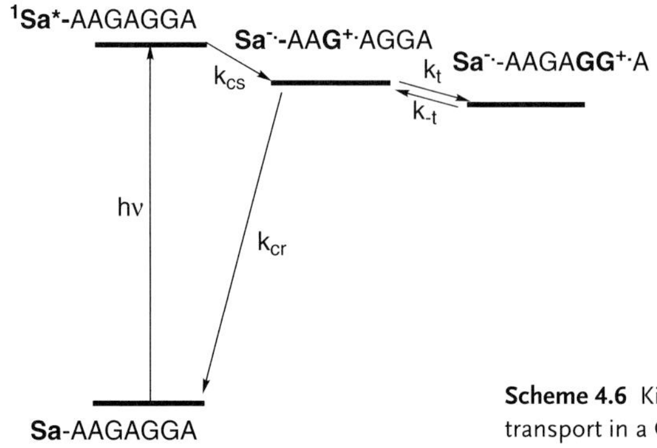

Scheme 4.6 Kinetic scheme for reversible hole transport in a GAGG sequence.

Table 4.1 Dynamics and equilibria for reversible hole transport in DNA.

	Sequence[a)]	$10^7\ k_t$, s^{-1}	$10^7\ k_{-t}$, s^{-1}	K_{ht} [b)]	$-\Delta G_{ht}$ (eV) [c)]
24	Sa-AAGAGGA	5.6	0.75	7.5	0.052
25	Sa-AAAGAGG	1.3	0.9	1.4	0.01
26	Sa-TTGAGGA	10.2	1.2	8.5	0.055
27	Sa-TTTGAGG	0.84	0.54	1.6	0.01
28	Sa-AAAGTGGA	0.4	0.06	6.7	0.49
29	Sa-AAAGACC[d)]	0.21	0.037	5.7	0.04
30	Sa-TTTGACC[d)]	0.10	0.033	3.0	0.03
31	Sa-AAGAGGGA	8.7	0.43	20	0.077
32	Sa-AAGAZAA	69	0.04	1700	0.19
33	Sa-AAGAAZA	5.0	<0.1	>500	>0.16

a) All sequences have 5′-3′ polarity. The complementary hairpin arm is not shown.
b) $K_{ht} = k_t/k_{-t}$.
c) $\Delta G_{ht} = -RT(\ln K_{ht})$.
d) GG secondary hole acceptor is located in the complementary strand.

sequence of hairpin **20** (see Figure 4.12). This difference may be a consequence of the larger solvent reorganization energy for a charge shift process vs. a charge separation process. Second, values of k_t for GAGG sequences **25** and **27** are slower than those for **24** and **26**. This difference can be attributed to the presence in **24** and **26** of a terminal A, which serves to stabilize a hole on the adjacent GG. Third, the value of k_t for the GTGGA sequence in **28** is ca. 20 ± 5 times slower than for **24** or **26**. This difference is consistent with the higher energy of the bridge orbitals for T vs. A. Fourth, the values of k_t for the GACC sequences in **30** and **31** are ca. 10 ± 5 times slower than for **25** or **27**. This difference represents the kinetic penalty for cross-strand hopping.

A change in the secondary hole donor from GG to GGG results in a modest increase in k_t for the GAGGGA hole transport sequence of **31** vs. **24**. A much larger increase in k_t is observed for the GAZA sequence in **32** vs. **24**. This is indicative of significantly more exothermic hole transport for Z vs. GG as the secondary acceptor. As previously noted, the oxidation potentials of the nucleobases within duplex DNA have not been determined experimentally. The single-nucleoside oxidation potential of Z is substantially lower than that of G (Scheme 4.3) [54]. Our data suggest that the oxidation potential of GG within duplex DNA is much closer to that of G than Z.

Finally, the introduction of a second A-T base pair in the hole transport sequence GAAZA results in a decrease in the value of k_t for **33** vs. **32** by a factor of 14. This factor is similar to that for hole injection in a **Sa**-AAG vs. **Sa**-AG sequence (Figure 4.12, **21** vs. **20**). The rates of hole transport in GAAGGA sequences are also much slower than those for GAGGA sequences. However, the long-lived transient signals for hairpins possessing GAAGGA sequences are too weak to permit kinetic modeling. Experimental studies of the slow hole transport dynamics over longer bridging sequences have been reported by the groups of Kawai and Majima and Shafirovich and Geacintov [55, 56].

4.4.3
Hole Transport Equilibria

The availability of rate constants for both forward and return charge transport from kinetic modeling permits calculation of the equilibrium constants and free energies of hole transport (Table 4.1). Values of $K_{ht} \sim 8$ are obtained for both GAGGA and GTGGA hole transport sequences in **24**, **26**, and **28**. Smaller values of $K_{ht} \sim 1.5$ are obtained for hairpins **25** and **27**, which have GAGG hole transport sequences. This suggests that the energy of hole on a terminal GG is similar to that of an internal G. A somewhat larger value of $K_{ht} = 20$ is obtained for the GAGGGA sequence of **31** and a much larger value of $K_{ht} = 1700$ for the GAZA sequence of **32**. The calculated free energy of hole transport in **32** is $\Delta G_{ht} = 0.19$ eV, approximately half the size of the reported difference in single-nucleoside oxidation potentials of Z vs. G (Scheme 4.3). The smaller value of ΔG_{ht} obtained from K_{ht} for **32** may indicate that interactions with adjacent bases serve to have a leveling effect on the single-base redox potentials.

The values of ΔG_{ht} for hole transport from G to GG or GGG in hairpins **24** and **31**, respectively, are 52 meV and 77 meV (1.2 kcal and 1.8 kcal), significantly smaller than the value for hole transport from G to Z. These values are also much smaller than the differences in calculated gas-phase ionization potentials for G vs. GG vs. GGG [46]. Recently, Kurnikov et al. reported that consideration of solvation greatly reduces the calculated difference in ionization potentials, with the solvation energy decreasing as the hole becomes more delocalized [57]. Our experimental values indicate that GG and GGG are very shallow "hole traps," in agreement with the ability of hole migration over multiple GG and GGG sites to compete with strand cleavage.

4.4.4
Hole Transport and Strand Cleavage

Most investigations of hole transport in DNA have employed strand cleavage efficiencies to determine relative rate constants for hole transport [12, 14, 58]. Strand cleavage patterns presumably are determined by the relative rate constants for hole transport and strand cleavage. A comparison of experimental strand cleavage results and our studies of hole transport dynamics and equilibrium has recently been published [59]. Observations of comparable yields of strand cleavage at multiple GG or GGG steps require that hole transport be more rapid than the chemical reactions that lead to strand cleavage. Relative yields of cleavage at G, GG, and GGG sites are consistent with the modest decrease in reactivity per G that accompanies the modest increase in hole stability.

To our knowledge, there is no single system for which the dynamics of both hole transport and stand cleavage efficiencies has been determined. We have attempted to study strand cleavage in several **Sa**-linked hairpins that possess both primary and secondary electron donors (Table 4.1). Unfortunately, the formation of base-labile products does not compete effectively with bleaching of the **Sa** chromophore. Continuous exposure of these hairpins to UV light results in a combination of stilbene photoisomerization [25] and perhaps also [2 + 2] addition of stilbene and thymine.

4.5
Electron Injection

4.5.1
Electron Injection to Neighboring Base Pairs

The energetics of photoinduced electron transfer from the singlet stilbenediether linker **Sd** to an adjacent nucleobase can be calculated using Eq. (2) and the data in Schemes 4.2 and 4.3. As shown in Scheme 4.4 b, photoreduction of T (and C) is expected to be exergonic, reduction of G endergonic, and reduction of A approximately isoergonic. We have investigated the dynamics of charge separation and charge recombination for the **Sd**-linked hairpins **34–39**, which possess different adjacent base pairs (Figure 4.13) [24, 60]. In all cases, the **Sd** singlet-state decay times (τ_s) are no longer than a few picoseconds. As shown in Figure 4.4, decay of the **Sd** singlet 575-nm band is accompanied by the appearance of a narrower band at 525 nm assigned to **Sd**$^{+\bullet}$. The decay of the latter band is attributed to charge recombination of the radical ion pair and has decay times (τ_{rip}) between 10 ps and 40 ps. Values of k_{cs} and k_{cr} calculated from the measured decay times are shown in Figure 4.13.

The values of k_{cs} and k_{cr} can be correlated with the reduction potential or electron affinity of the adjacent nucleobases (Scheme 4.3), which follow the order BrU < T < C < A < G. A decrease in reduction potential results in more exergonic

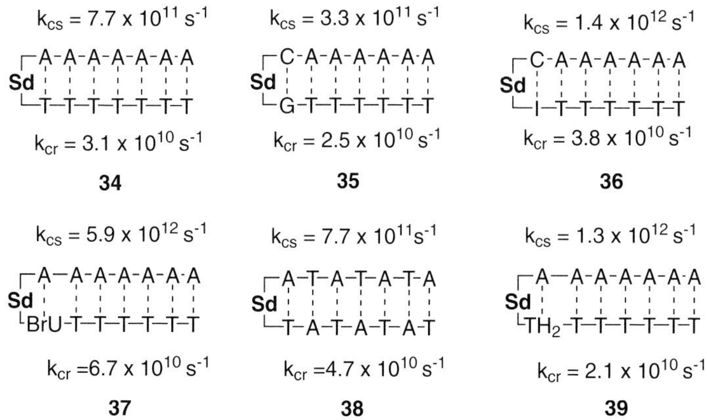

Fig. 4.13 Rate constants for charge separation and charge recombination in some **Sd**-linked hairpins.

charge separation, leading to an increase in k_{cs}. A decrease in reduction potential will also result in less exergonic charge recombination (Eq. 3). For charge recombination in the Marcus inverted region, the rate constant k_{cr} should increase as ΔG_{cr} becomes smaller (Figure 4.9).

Brédas and coworkers recently predicted the occurrence of strong electronic coupling between singlet **Sd** and the band-like electronic structure of an adjacent poly(A-T) sequence [43]. We reasoned that replacement of a poly(A) strand with an alternating $(AT)_3$ sequence should destroy any poly(A) band-like structure. In fact, the values of k_{cs} are similar for hairpins **34** and **38**. The value of k_{cr} is slower for **34**, which might indicate somewhat greater delocalization of the radical ion pair or stabilization of $A^{-\bullet}$ by an adjacent A.

To our knowledge the reduced thymine derivative TH_2 (Scheme 4.3) has not previously been employed in studies of DNA electron transfer. Incorporation of TH_2 into duplex DNA is reported to occur with minimal perturbation of duplex structure or function [61]. We anticipated that it would be more difficult to reduce TH_2 than T as a consequence of loss of conjugation. However, the values of τ_s and τ_{cr} for **Sd**-linked hairpins **39** and **34** containing adjacent TH_2-A and T-A base pairs are similar. Plausibly, the anion radicals of T and TH_2 may be localized mainly on the imide portion of the molecule and thus may not be sensitive to the presence or absence of the C=C bond.

In the case of the BrU-containing hairpin **37**, there is a long-lived decay component (>2 ns) that accounts for ca. 5% of the total $Sd^{+\bullet}$ decay. It is tempting to assign this persistent $Sd^{+\bullet}$ signal to hairpins in which loss of halide ion competes with charge recombination. Studies of halide loss and strand cleavage designed to explore this possibility are in progress.

4.5.2
Electron Injection via a GG Base Pair

Electron injection from singlet **Sd** to G is expected to be endergonic (Scheme 4.4 b). Separation of the **Sd** linker from bases of low reduction potential such as T, C, or BrU by a G-G base pair thus might be expected to slow the rate of charge separation, resulting in an increase in τ_S. G-G base pairs are among the most commonly encountered mismatched base pairs [62]. They distort the sugar geometry, leading to a decrease in duplex stability, but do not distort the base stacking of duplex DNA. The values of τ_S for hairpins **40–44** (Figure 4.14) are in accord with this expectation. The ratio of τ_S values for **40** vs. **34** is ca. 130 : 1, much larger than the decrease in the rate constants for hole injection in **Sa**-linked hairpins **4** vs. **3** (Figure 4.4). Addition of a second G-G base pair results in an even larger value of τ_S for hairpin **41**. Assuming that the G-G base pairs do not distort the hairpin structure, these results are indicative of a much steeper distance dependence for electron injection vs. hole injection. This conclusion is consistent with the results of pulse radiolysis studies of DNA electron injection [63].

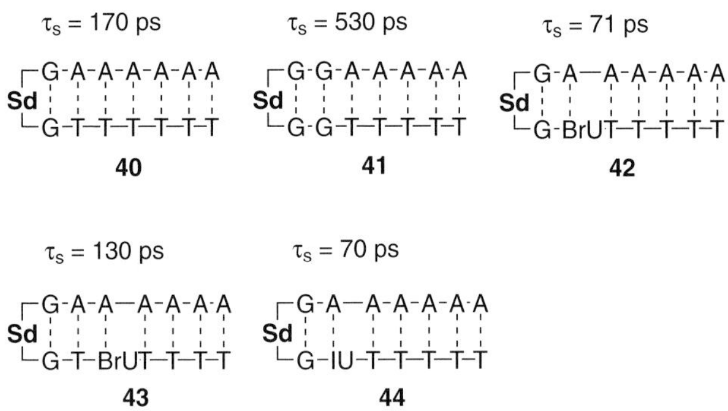

Fig. 4.14 Singlet-state decay times for some **Sd**-linked hairpins.

Hairpins **42–44** have shorter singlet lifetimes than does **40**, which is indicative of more-effective long-range electron injection for BuU or IU vs. T acceptors. The transient spectra of **40–44** are dominated by the broad absorption of singlet **Sd**, which overlaps the narrower transient absorption of $\mathbf{Sd}^{+\bullet}$ (Figure 4.3). This prevents analysis of the $\mathbf{Sd}^{+\bullet}$ rise and decay times. However, there is a long-lived component of $\mathbf{Sd}^{+\bullet}$ decay that persists beyond the 2–4 nanosecond limit of the picosecond laser experiment. This component accounts for ca. 10 % of the total 525-nm decay of hairpins **42–44**, but <5 % of the 525-nm decay of hairpin **34**. As suggested in the case of hairpin **37**, this component may result from loss of halide ions from the BuU anion radical.

4.5.3
Electron Injection and Strand Cleavage

Several groups have used chemical assays as a means of studying the distance dependence of electron injection in duplex DNA [64, 65]. The two assays employed are cleavage of T-T dimers [66–68] and strand cleavage resulting in loss of halide ions from reduced BrU or IU [63, 69]. A shallow distance dependence for relative yields of reaction has been reported for some, but not all, of these systems. In the absence of quantum yield data, the efficiencies of these processes, as well as their dynamics, remain undetermined.

Acknowledgments

We thank our coworkers for their contributions to these studies and our colleague Robert Letsinger for engaging our interest in DNA. This research was supported by grants from the Division of Chemical Sciences, Office of Basic Energy Sciences, U.S. Department of Energy under contracts DE-FG02-96ER14604 (FDL) and DE-FG02-99ER14999 (MRW).

References

1 S. Neidle, *Oxford Handbook of Nucleic Acid Structure,* Oxford University Press, Oxford, **1999**.
2 V. A. Bloomfield, D. M. Crothers, I. Tinoco, Jr., *Nucleic Acids, Structures, Properties, Functions,* University Science Books, Sausalito, CA, **2000**.
3 C. R. Cantor, P. R. Schimmel, *Biophysical Chemistry,* Vol. 2, W. H. Freeman, New York, **1980**.
4 D. D. Eley, D. I. T. Spivey, *Trans. Faraday Soc.* **1962**, *58*, 411–415.
5 J. D. Watson, F. H. C. Crick, *Nature* **1953**, *171*, 737–738.
6 M. P. O'Neil, E. M. Fieldew, *Adv. Radiation Biol.* **1993**, *17*, 53–120.
7 C. J. Murphy, M. R. Arkin, Y. Jenkins, N. D. Ghatlia, S. H. Bossmann, N. J. Turro, J. K. Barton, *Science* **1993**, *262*, 1025–1029.
8 C. J. Murphy, M. R. Arkin, N. D. Ghatlia, S. Bossmann, N. J. Turro, J. K. Barton, *Proc. Natl. Acad. Sci. USA* **1994**, *91*, 5315–5319.
9 M. R. Arkin, E. D. A. Stemp, R. E. Holmlin , J. K. Barton, A. Hörmann, E. J. C. Olson, P. F. Barbara, *Science* **1996**, *273*, 475–479.
10 F. D. Lewis, in *Electron Transfer in Chemistry, Vol. 3* (Ed.: V. Balzani), Wiley-VCH, Weinheim, Germany, **2001**.
11 P. F. Barbara, E. J. C. Olson, *Adv. Chem. Phys.* **1999**, *107*, 647–676.
12 B. Giese, *Acc. Chem. Res.* **2000**, *33*, 631–636.
13 B. Giese, A. Biland, *Chem. Commun.* **2002**, 667–672.

14 G. B. Schuster, *Acc. Chem. Res.* **2000**, *33*, 253–260.
15 E. Tuite, in *Organic and Inorganic Photochemistry, Vol. 2* (Eds.: V. Ramamurthy, K. S. Schanze), Marcel Dekker, New York, **1998**, pp. 55–74.
16 R. E. Holmlin, P. J. Dandliker, J. K. Barton, *Angew. Chem. Int. Ed. Engl.* **1997**, *36*, 2714–2730.
17 E. K. Wilson, *Chem. Eng. News* **1998**, *75*, 51–54.
18 G. B. Schuster, *Long-Range Charge Transfer in DNA*, Springer-Verlag, Berlin, **2004**.
19 F. D. Lewis, T. Wu, Y. Zhang, R. L. Letsinger, S. R. Greenfield, M. R. Wasielewski, *Science* **1997**, *277*, 673–676.
20 F. D. Lewis, T. Wu, X. Liu, R. L. Letsinger, S. R. Greenfield, S. E. Miller, M. R. Wasielewski, *J. Am. Chem. Soc.* **2000**, *122*, 2889–2902.
21 F. D. Lewis, R. S. Kalgutkar, Y. Wu, X. Liu, J. Liu, R. T. Hayes, M. R. Wasielewski, *J. Am. Chem. Soc.* **2000**, *122*, 12346–12351.
22 F. D. Lewis, J. Liu, W. Weigel, W. Rettig, I. V. Kurnikov, D. N. Beratan, *Proc. Natl. Acad. Sci. U.S.A.* **2002**, *99*, 12536–12541.
23 F. D. Lewis, J. Liu, X. Zuo, R. T. Hayes, M. R. Wasielewski, *J. Am. Chem. Soc.* **2003**, *125*, 4850–4861.
24 F. D. Lewis, X. Liu, S. E. Miller, R. T. Hayes, M. R. Wasielewski, *J. Am. Chem. Soc.* **2002**, *124*, 11280–11281.
25 R. L. Letsinger, T. Wu, *J. Am. Chem. Soc.* **1995**, *117*, 7323–7328.
26 F. D. Lewis, Y. Wu, X. Liu, *J. Am. Chem. Soc.* **2002**, *124*, 12165–12173.
27 F. D. Lewis, X. Liu, Y. Wu, S. E. Miller, M. R. Wasielewski, R. L. Letsinger, R. Sanishvili, A. Joachimiak, V. Tereshko, M. Egli, *J. Am. Chem. Soc.* **1999**, *121*, 9905–9906.
28 M. Egli, V. Tereshko, R. Mushudov, R. Sanishvili, X. Liu, F. D. Lewis, *J. Am. Chem. Soc.* **2003**, *125*, 10842–10849.
29 A. Weller, *Zeit. Phys. Chem. Neue. Folg.* **1982**, *133*, 93–98.
30 C. A. M. Seidel, A. Schulz, M. H. M. Sauer, *J. Phys. Chem.* **1996**, *100*, 5541–5553.
31 F. D. Lewis, R. L. Letsinger, M. R. Wasielewski, *Acc. Chem. Res.* **2001**, *34*, 159–170.
32 J. R. Winkler, H. B. Gray, *Chem. Rev.* **1992**, *92*, 369–379.
33 W. B. Davis, W. A. Svec, M. A. Ratner, M. R. Wasielewski, *Nature* **1998**, *396*, 60–63.
34 C. Wan, T. Fiebig, O. Schiemann, J. K. Barton, A. H. Zewail, *Proc. Natl. Acad. Sci. U.S.A.* **2000**, *97*, 14052–14055.
35 K. Fukui, T. Tanaka, *Angew. Chem. Int. Ed.* **1998**, *37*, 158–161.
36 S. Hess, M. Götz, W. B. Davis, M. E. Michel-Beyerle, *J. Am. Chem. Soc.* **2001**, *123*, 10046–10055.
37 A. Harriman, *Angew. Chem. Int. Ed.* **1999**, *38*, 945–949.
38 A. M. Brun, A. Harriman, *J. Am. Chem. Soc.* **1992**, *114*, 3656–3660.
39 E. J. C. Olson, D. Hu, A. Hörmann, P. F. Barbara, *J. Phys. Chem.* **1997**, *101*, 299–303.

40 F. D. Lewis, Y. Wu, L. Zhang, X. Zuo, R. T. Hayes, M. R. Wasielewski, *J. Am. Chem. Soc.* **2004**, *126*, 8206–8215.

41 F. C. Grozema, Y. A. Berlin, L. D. A. Siebbeles, *Int. J. Quantum Chem.* **1999**, *75*, 1009–1016.

42 H. L. Tavernier, M. D. Fayer, *J. Phys. Chem. B* **2000**, *104*, 11541–11550.

43 D. Beljonne, G. Pourtois, M. A. Ratner, J. L. Brédas, *J. Am. Chem. Soc.* **2003**, *125*, 14510–14517.

44 D. N. LeBard, M. Lilichenko, D. V. Matyushov, Y. A. Berlin, M. A. Ratner, *J. Phys. Chem. B.* **2003**, *107*, 14509–14520.

45 B. Armitage, *Chem. Rev.* **1998**, *98*, 1171–1200.

46 H. Sugiyama, I. Saito, *J. Am. Chem. Soc.* **1996**, *118*, 7063–7068.

47 M. F. Sistare, S. J. Codden, G. Heimlich, H. H. Thorp, *J. Am. Chem. Soc.* **2000**, *122*, 4742–4749.

48 F. D. Lewis, J. Liu, X. Liu, X. Zuo, R. T. Hayes, M. R. Wasielewski, *Angew. Chem. Int. Ed.* **2002**, *41*, 1026–1028.

49 W. B. Davis, I. Naydenova, R. Haselberger, A. Ogrodnik, B. Giese, M. E. Michel-Beyerle, *Angew. Chem. Int. Ed.* **2000**, *39*, 3649–3652.

50 F. D. Lewis, X. Liu, S. E. Miller, R. T. Hayes, M. R. Wasielewski, *J. Am. Chem. Soc.* **2002**, *124*, 14020–14026.

51 F. D. Lewis, X. Liu, J. Liu, S. E. Miller, R. T. Hayes, M. R. Wasielewski, *Nature* **2000**, *406*, 51–53.

52 F. D. Lewis, X. Liu, J. Liu, R. T. Hayes, M. R. Wasielewski, *J. Am. Chem. Soc.* **2000**, *122*, 12037–12038.

53 F. D. Lewis, X. Zuo, J. Liu, R. T. Hayes, M. R. Wasielewski, *J. Am. Chem. Soc.* **2002**, *124*, 4568–4569.

54 S. O. Kelley, J. K. Barton, *Science* **1999**, *283*, 375–381.

55 K. Kawai, T. Takada, S. Tojo, N. Ichinose, T. Majima, *J. Am. Chem. Soc.* **2001**, *123*, 12688–12689.

56 V. Y. Shafirovich, A. Dourandin, W. Huang, N. P. Luneva, N. E. Geacintov, *Phys. Chem. Chem. Phys.* **2000**, *2*, 4399–4408.

57 I. V. Kurnikov, G. S. M. Tong, M. Madrid, D. N. Beratan, *J. Phys. Chem. B* **2002**, *106*, 7–10.

58 M. E. Núñez, D. B. Hall, J. K. Barton, *Chem. & Biol.* **1999**, *6*, 85–97.

59 F. D. Lewis, M. R. Wasielewski, *Top. Cur. Chem.* **2004**, *236*, 45–65.

60 F. D. Lewis, X. Liu, H. Zhu, R. T. Hayes, L. E. Sinks, *unpublished results.*

61 K. Miaskiewicz, J. Miller, R. Ornstein, R. Osman, *Biopolymers* **1995**, *35*, 113–124.

62 M. E. Burkard, D. H. Turner, *Biochemistry* **2000**, *39*, 11748–11762.

63 K. Kawai, T. Kimura, K. Kawabata, S. Tojo, T. Majima, *J. Phys. Chem. B.* **2003**, *107*, 12838–12841.

64 C. Behrens, M. K. Cichon, F. Grolle, U. Hennecke, T. Carell, *Top. Cur. Chem.* **2004**, *236*, 187–204.

65 T. Carell, C. Behrens, J. Gierlich, *Org. Biomol. Chem.* **2003**, *1*, 2221–2228.

66 T. Ito, S. E. Rokita, *J. Am. Chem. Soc.* **2003**, *125*, 11480–11481.

67 S. Breeger, U. Hennecke, T. Carell, *J. Am. Chem. Soc.* **2004**, *126*, 1302–1303.
68 B. Giese, B. Carl, T. Carl, T. Carell, C. Behrens, U. Hennecke, O. Schiemann, E. Feresin, *Angew. Chem. Int. Ed.* **2004**, *43*, 1848–1851.
69 T. Oyoshi, A. H.-J. Wang, H. Sugiyama, *J. Am. Chem. Soc.* **2002**, *124*, 2086–2087.

5
Spectroscopic Investigation of Oxidative Hole Transfer via Adenine Hopping in DNA

Kiyohiko Kawai and Tetsuro Majima

5.1
Introduction

One-electron oxidation of DNA leads to formation of the radical cation of guanine ($G^{\bullet +}$), with the lowest oxidation potential among the four bases, and a hole has been demonstrated to migrate through DNA over a long distance (~200 Å) by hopping between G's [1–4]. The rate constant for a single-step charge transfer process (k_{ct}) usually follows an exponential dependence on the donor-acceptor distance Δr:

$$\ln k_{ct} \propto -\beta \Delta r \quad (1)$$

From the spectroscopic measurements of the charge transfer in DNA, the distance dependence parameter β in Eq. (1) has been determined as 0.6 $Å^{-1}$ [5–7]. However, strand cleavage experiments revealed the occurrence of the long-range hole transfer in DNA containing long intervening A-T sequences between G's, where, judging from the β-value of 0.6 $Å^{-1}$, hole transfer between G's should be too slow to compete with the hole-trapping reactions [2, 3]. Giese et al. explained this contradiction by showing that hole transfer can be mediated by thermally induced hopping between the second most easily oxidized base adenine (A-hopping) [8]. Thus, hole transfer in DNA has been demonstrated to occur by the two different mechanisms, G-hopping (superexchange between G's across the intervening A-T bridge) and A-hopping (charge is carried by the bridge base A as the A radical cation [$A^{\bullet +}$]). The rate constants of the hole transfer by G-hopping across one A-T base pair have been determined by Lewis et al. to be in the range of 10^6–10^8 s^{-1} [9]. Considering the low efficient endothermic oxidation of A by $G^{\bullet +}$ and the weak distance dependence of A-hopping, it was suggested that after the generation of $A^{\bullet +}$, the hopping between A proceeds quickly [10]. However, little was known about the kinetics of hole transfer by A-hopping. In this chapter, we describe our research on the spectroscopic measurements, especially addressing the kinetics of hole transfer by A-hopping in DNA.

Charge Transfer in DNA: From Mechanism to Application. Edited by H.-A. Wagenknecht

ISBN: 3-527-31085-1

5.2 Kinetics of Hole Transfer in DNA by Adenine Hopping

5.2.1 Pulse Radiolysis–laser Flash Photolysis of Phenothiazine-modified ODN

For the measurement of the hole transfer rates in DNA, it is desirable to generate a hole with a sufficiently long lifetime in DNA. For this purpose, pulse radiolysis serves as a suitable method in which a hole is irreversibly generated in DNA free from charge recombination. We have previously reported the pulse radiolysis of pyrene (Py)-conjugated and phenothiazine (PTZ)-conjugated oligodeoxynucleotides (ODNs), in which hole transfer in DNA was monitored by the transient absorption of radical cations of Py and PTZ ($Py^{\bullet+}$ and $PTZ^{\bullet+}$), respectively [7, 11–14]. However, pulse radiolysis is disadvantageous in that it cannot be used to measure the kinetics on a time scale shorter than 1 μs because of the time taken for the collisional process for the generation of a hole in DNA. Furthermore, collisional oxidation of DNA makes it difficult to generate a hole at the desired position in DNA. Here, to avoid these difficulties, we employed the pulse radiolysis–laser flash photolysis of PTZ-modified ODN (PTZ-ODNn). First, electron pulse (28 MeV, 8 ns) was applied to the N_2O-saturated aqueous solution containing 2 mM Tl_2SO_4, 20 mM pH 7.0 Na phosphate buffer, and 0.2 mM (strand conc.) PTZ-ODNn to generate the oxidizing reagent $TlOH^+$. After the collisional oxidation of PTZ-ODNn by $TlOH^+$, i.e., after the generation of $PTZ^{\bullet+}$ with a maximum absorption peak at 520 nm, $PTZ^{\bullet+}$ was irradiated with a 532-nm laser flash to produce $PTZ^{\bullet+}$ in the excited state ($PTZ^{\bullet+*}$). The energy of $PTZ^{\bullet+*}$ was estimated to be higher than 2 eV from the edge of the absorption peak of $PTZ^{\bullet+}$ at the longer wavelength side. Using the oxidation potential of PTZ ($E_{ox} = 0.76$ V vs. NHE in CH_3CN [12]), the reduction potential of $PTZ^{\bullet+*}$ was estimated to be ~2.8 eV. Thus, the hole transfer from $PTZ^{\bullet+*}$ to A ($E_{ox} = 1.42$ V vs. NHE in H_2O [15]; $E_{ox} = 1.96$ V vs. NHE in CH_3CN [16]) will be exergonic, and $A^{\bullet+}$ will be mainly produced when A is the nearest base to the $PTZ^{\bullet+*}$ [17]. We used 8-oxo-7,8-dihydroguanine (oxG) as a hole trap to avoid further hole transfer between G's along DNA. Hence, pulse radiolysis–laser flash photolysis of PTZ-ODNn leads to the selective injection of a hole at A nearest to PTZ, and the forward hole transfer from $A^{\bullet+}$ across $(A)_n$ bridge to oxG and the backward hole transfer from the radical cation of oxG ($oxG^{\bullet+}$ or $oxG(-H^+)^{\bullet}$:$C(+H^+)^+$) to PTZ were investigated (Figure 5.1).

Irradiation of $PTZ^{\bullet+}$ in PTZ-ODN1 with the 532-nm laser flash with a delay time of 50 μs to the electron pulse caused a decrease in ΔOD of $PTZ^{\bullet+}$ and a formation of broad absorption around 400 nm immediately after the flash (Figures 5.2 and 5.3, see p. 120). The new absorption band was assigned to $oxG^{\bullet+}$ [18], demonstrating the hole transfer from $PTZ^{\bullet+*}$ to oxG (forward hole transfer) within the laser flash duration of 5 ns, i.e., the forward hole transfer proceeds faster than 10^8 s^{-1}. The spectrum of $PTZ^{\bullet+}$ was recovered in the time scale of ~500 μs concomitant with the decay of $oxG^{\bullet+}$ according to the backward hole transfer from $oxG^{\bullet+}$ to PTZ. Bleaching of the transient absorption at 520 nm assigned to $PTZ^{\bullet+}$

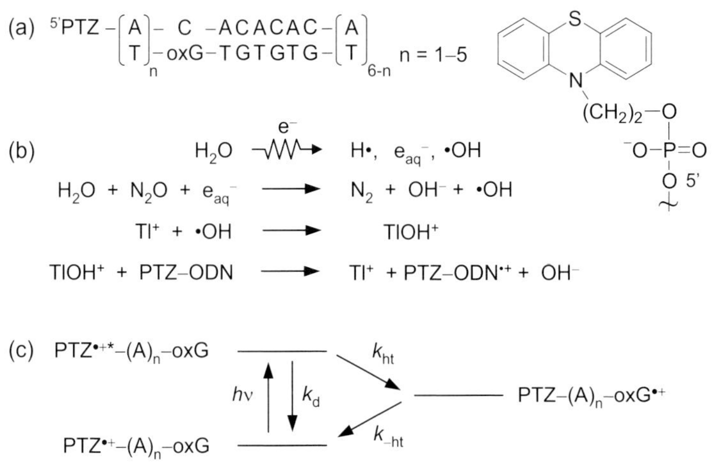

Fig. 5.1 (a) Sequence and chemical structure of phenothiazine (PTZ)-modified ODNs (PTZ-ODNn). (b) Kinetic scheme for generation of oxidizing reagent $TlOH^+$ and one-electron oxidation of PTZ-ODNn during pulse radiolysis. (c) Kinetic scheme for laser-stimulated forward (k_{ht}) and backward (k_{-ht}) hole transfer.

upon the laser flash ($\Delta\Delta OD_{520}$) weakly decreased as the distance between the PTZ and oxG increased, demonstrating the weak influence of the distance on the forward hole transfer rate. In contrast, backward hole transfer from $oxG^{\bullet+}$ to PTZ was strongly distance dependent, and when PTZ and $oxG^{\bullet+}$ were separated by more than three A-T base pairs, no backward hole transfer was observed in the present experimental time scale. Thus, different distance dependence was observed in the forward and backward hole transfers in PTZ-ODNn.

5.2.2
Hole Transfer by Superexchange and Hopping

The relative rate for the forward hole transfer (k_{ht}) follows Eq. (2), where Φ and k_d designate the yield of the forward hole transfer ($\Phi = \Delta\Delta OD_{520}/\Delta OD_{520\ \text{before laser}}$) and the rate of deactivation of $PTZ^{\bullet+*}$, respectively. Then, ln k_{ht} can be described by Eq. (3).

$$\Phi = k_{ht}/(k_{ht} + k_d) \tag{2}$$

$$\ln k_{ht} = \ln(\Phi/(1-\Phi)) + \text{const.} \tag{3}$$

The rate of the backward hole transfer (k_{-ht}) was directly determined from the recovery of $PTZ^{\bullet+}$. Figure 5.4a shows the semi-log plot of $\Phi/(1-\Phi)$ and k_{-ht} against the distance between the PTZ and oxG (Δr). From the slope of the plot, values of 0.2 $Å^{-1}$ and 0.6 $Å^{-1}$ were obtained for the distance dependence of k_{ht} and k_{-ht}, respectively. The β-value of 0.6 $Å^{-1}$ obtained for the backward hole transfer is

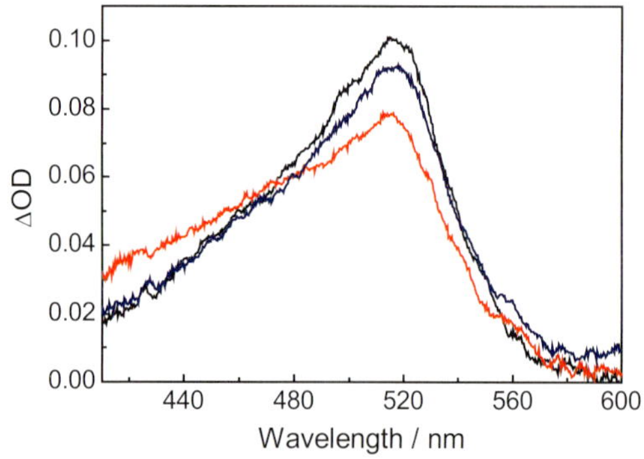

Fig. 5.2 Transient absorption spectra of PTZ-ODN1 obtained at 45 μs (black), 55 μs (red), and 500 μs (blue) after the electron pulse during pulse radiolysis–laser flash photolysis. The 532-nm laser flash was delayed by 50 μs relative to the electron pulse.

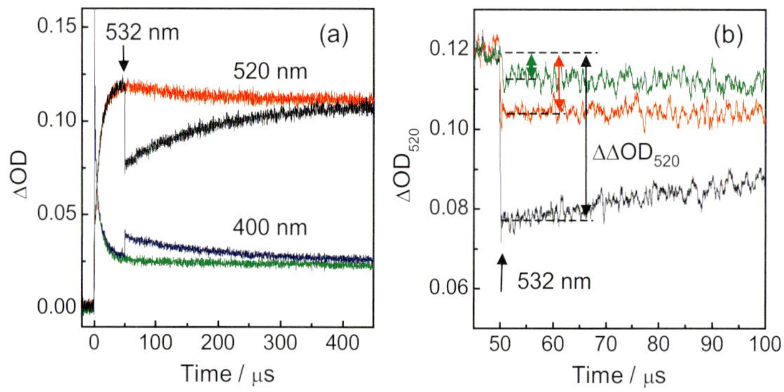

Fig. 5.3 (a) Time profiles of transient absorptions monitored at 520 nm (black, red: without laser flash) and 400 nm (blue, green: without laser flash) during the pulse radiolysis–laser flash photolysis of PTZ-ODN1. The delay time of the laser flash was 50 μs after the electron pulse. (b) Bleaching of $PTZ^{\bullet+}$ monitored at 520 nm upon the irradiation of the 532-nm laser flash for PTZ-ODNn (black, n = 1; red, n = 3; green, n = 5).

consistent with the single-step charge transfer in DNA. In contrast, the value of 0.2 $Å^{-1}$ for the forward hole transfer is significantly smaller than the reported β-values for the single-step charge transfer process [19].

The distance dependence of the charge transfer by the hopping mechanism can be described in the simplest way by Eq. (4), in which η is a proportional factor and N is the number of hopping steps [20–22]. In the case of random walk, η takes the value between $1 \leq \eta \leq 2$.

$$\ln k \propto -\eta \ln N \tag{4}$$

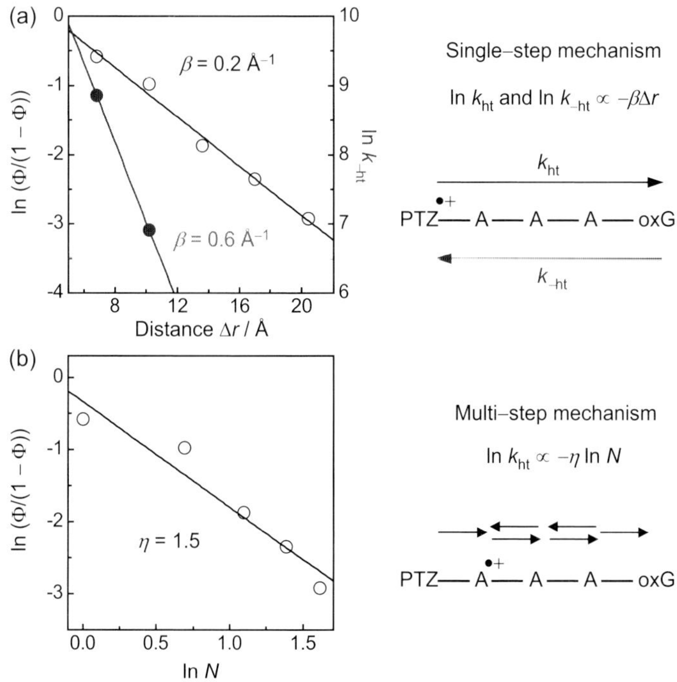

Fig. 5.4 Dependence of the forward and backward hole transfer rates (o and •, respectively) (a) on the distance between PTZ and oxG (Δr) and (b) on the number of hopping steps (N).

A plot of ln $(\Phi/(1-\Phi))$ against ln N gave a straight line and an η-value of 1.5 (Figure 5.4b). Thus, the forward hole transfer was demonstrated to proceed by A-hopping rather than by the single-step superexchange mechanism.

In this section, two different processes for the hole transfer across the identical $(A)_n$ bridge in DNA have been demonstrated. Since $PTZ^{\bullet+*}$ oxidizes A, $PTZ^{\bullet+*}$ injects a hole at A nearest to PTZ, and the forward hole transfer proceeds by the A-hopping mechanism. However, $oxG^{\bullet+}$ cannot oxidize A, resulting in hole transfer by the single-step superexchange mechanism across the $(A)_n$ bridge. These results clearly show that the mechanism of hole transfer in DNA strongly depends on the redox nature of the oxidant, whether it produces only $G^{\bullet+}$ or both $A^{\bullet+}$ and $G^{\bullet+}$. Hole transfer in DNA by A-hopping was demonstrated to proceed faster than 10^8 s^{-1} over the distance range of 7–22 Å. Hence, the fast hole transfer rates reported in the previous literature [23] may be explained by the hole transfer by A-hopping.

$$\Phi' = \frac{\Phi_{cs}}{1 - \Phi_{cs}} = \frac{(k_2/k_1)}{1 + (N-1)(k_2/k_A)} \tag{6}$$

where k_A, k_1, and k_2 are the rate constants of the A-hopping, charge recombination between $NDI^{\bullet-}$ and $A^{\bullet+}$, and the hole shift from adjacent $A^{\bullet+}$ to PTZ, respectively. The electron transfer process from $^1NDI^*$ to nucleobases and charge recombination between the nucleobase radical cation and $NDI^{\bullet-}$ were previously reported by Lewis and coworkers, demonstrating that the charge separation and recombination rates between NDI and the A-T pair were $7.3 \times 10^{12}\ s^{-1}$ and $k_1 = 2.5 \times 10^{11}\ s^{-1}$, respectively [32]. The rate constants for the A-hopping process were obtained by fitting Φ_{cs} depending on the hopping number (N) to Eq. (6), using the k_1 value of $2.5 \times 10^{11}\ s^{-1}$ to provide the k_A of $10^{10}\ s^{-1}$ (Figure 5.8). This value of k_A is much larger than that of hopping from $G^{\bullet+}$ to GG across one A base ($5 \times 10^7\ s^{-1}$) [9]. This difference is attributed to the shorter distance and the direct stacking between the A bases for the A-hopping presented in this study.

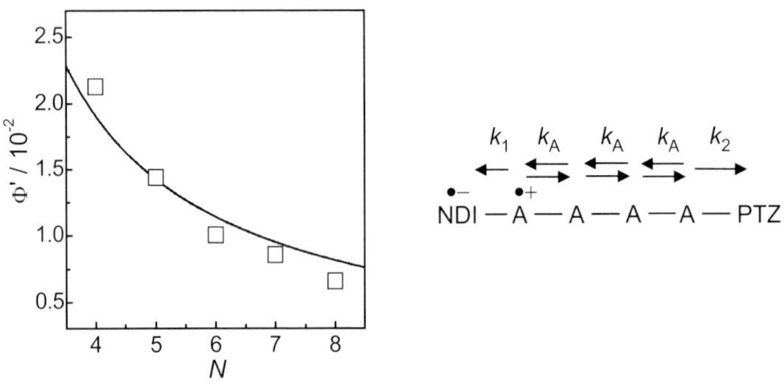

Fig. 5.8 Dependence of quantum yields ratio (Φ') for the charge separation upon the number of hopping steps (N).

In this section, it was demonstrated that a long-lived, charge-separated state in DNA can be achieved by using A-hopping. The rate constant of the A-hopping was determined to be $10^{10}\ s^{-1}$ by analyzing the obtained charge-separation yields. The charge-separation yield was approximately 2% or lower due to the rapid charge recombination between $NDI^{\bullet-}$ and adjacent $A^{\bullet+}$ in the contact ion pair. However, a high efficiency of the final charge-separated state in DNA may be achieved by replacing NDI with suitable acceptor molecules, such as triplet sensitizers, which can slow down the initial charge recombination.

5.4
Effect of Hole Transfer on Photosensitized DNA Damage

The one-electron oxidation of DNA occurs during photoirradiation in the presence of photosensitizers (Sens) and has been extensively studied because it leads to the formation of oxidative lesions that cause carcinogenesis and aging [41, 42]. Photosensitized DNA damage has also received attention from a therapeutic point of view, since DNA is one of the potential targets of photodynamic therapy [43]. The photoirradiation of DNA-bound Sens produces the radical anion of Sens ($Sens^{\bullet -}$) and $G^{\bullet +}$ as the charge-separated state through photoinduced electron transfer [44]. The efficiency of the photoinduced one-electron oxidation of DNA is seemingly low since the charge recombination rate is usually much faster than the process leading to the DNA strand cleavage, such as the reaction of $G^{\bullet +}$ with water [7, 9, 18]. However, photosensitized DNA damage does occur [45]. In the previous sections, we reported the kinetics of hole transfer in DNA by A-hopping, which is weakly distance dependent and proceeds faster than $10^8\ s^{-1}$ over the distance range of 7–22 Å [46, 47]. These findings prompted us to suggest that a fast hole transfer by A-hopping may help to separate the hole and $Sens^{\bullet -}$ during the photosensitized one-electron oxidation of DNA, providing the time for $G^{\bullet +}$ and $Sens^{\bullet -}$ to react with water or O_2. Here, to assess this hypothesis, laser flash photolysis and HPLC analysis of NDI-modified ODNs were performed.

NDI was selected as a Sens since $^1NDI^*$ can oxidize A to promote hole transfer by the A-hopping to eventually yield $G^{\bullet +}$ [33, 37, 47–49]. Several ODNs with different distances between the NDI and G's with intervening A_n sequences (NDn) were synthesized [47, 50] and the effect of the hole transfer on the DNA damage was investigated (Table 5.2). The excitation of NDI-modified ODN with the 355-nm laser (5 ns, 5 mJ $pulse^{-1}$) produced $NDI^{\bullet -}$ and $ODN^{\bullet +}$ in a charge-separated state through the photoinduced electron transfer, and the charge separation and recombination processes were examined by monitoring the formation and decay of $NDI^{\bullet -}$ as shown in Figure 5.9. In the case of NDn (n = 0–2) where G's are near the NDI, no transient absorption was observed due to the fast charge separation and charge recombination, which proceed within the laser flash duration of 5 ns. In NDn (n = 3–5) where the G's are separated from NDI by more than three base pairs, the formation of the transient absorption with a maximum peak at 495 nm was observed immediately after the flash excitation, which was assigned to $NDI^{\bullet -}$ [51]. The yield of the formed $NDI^{\bullet -}$ was similar for NDn (n = 3–5). In contrast, the lifetime of the charge-separated state significantly increased with the increasing of the distance between NDI and G's, i.e., the charge recombination process is strongly distance dependent. These results are consistent with the charge separation by the A-hopping and charge recombination by the superexchange mechanism as described in the previous sections [8, 10, 46, 52–54]. The charge separation yield is small, about 2% [47], owing to the fast charge recombination from the contact radical ion pair [55]. Therefore, the charge-separated state is generated by occasional escape from the charge recombination by the hole-shift process. However, once a hole escapes from the coulombic interaction, it effi-

Table 5.2 Decay lifetime of $NDI^{\bullet -}$ and consumption of G in photosensitized DNA damage.[a)]

ODNs	Sequence	τ (μs)[b)]	G (%)[c)]
ND0	5′NDI-CGCGCTTTTT 3′GCGCGAAAAA	<0.005	<0.1
ND1	5′NDI-TCGCGCTTTT 3′AGCGCGAAAA	<0.005	<0.1
ND2	5′NDI-TTCGCGCTTT 3′AAGCGCGAAA	<0.005	<0.1
ND3	5′NDI-TTTCGCGCTT 3′AAAGCGCGAA	0.24	0.36
ND4	5′NDI-TTTTCGCGCT 3′AAAAGCGCGA	0.20 (48), 7.1 (52)	2.6
ND5	5′NDI-TTTTTCGCGC 3′AAAAAGCGCG	6.0 (31), 60 (69)	10
NDG	5′NDI-TCTTTTGCGC 3′AGAAAACGCG	<0.005	1.5

a) Laser flash photolysis was carried out in an aqueous solution containing 40 μM ODN (strand conc.) and 20 mM pH 7.0 Na phosphate buffer.
b) The decay lifetime of $NDI^{\bullet -}$ (pre-exponential).
c) ODNs were photoirradiated with a 355-nm laser (1.6 mJ $pulse^{-1}$, 1500 pulses, total irradiated energy of 2.4 J) and digested with snake venom phosphodiesterase/nuclease P1/alkaline phosphatase to 2′-deoxyribonucleosides. The consumption of G was quantified by HPLC using A as an internal standard.

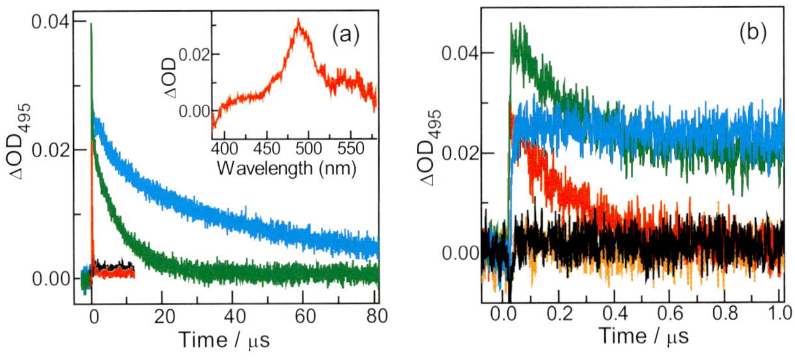

Fig. 5.9 Time profiles of the transient absorption of $NDI^{\bullet -}$ monitored at 495 nm during the 355-nm laser flash photolysis of Ar-saturated aqueous solution of NDI-modified ODNs: ND2 (orange), ND3 (red), ND4 (green), ND5 (cyan), and NDG (black) (a) in the time scale of 0–80 μs and (b) 0–1 μs. The inset shows the transient absorption spectrum of $NDI^{\bullet -}$ obtained at 100 ns after the 355-nm flash excitation of ND3.

ciently migrates through DNA by A-hopping [47]. Insertion of a single G in the A_n sequence between NDI and G's significantly diminished the transient absorption of $NDI^{\bullet-}$ (NDG), since the inserted G serves as a hole trap on the hole-shift process, causing inhibition of consecutive A-hopping in which a hole migrates to G's [56, 57].

To investigate the effect of the hole transfer on the DNA damage during the photosensitized one-electron oxidation of DNA, NDn was photoirradiated and the consumption of G was quantified by HPLC (Table 5.2). Interestingly, the consumption of G increased with the increasing of the distance between NDI and G's. In the case of NDG, where most of the generated holes recombine within the laser duration because of the inserted G in the A_n sequence, the consumption of G was small even though the remaining G's locate far from NDI. Thus, not the distance between the NDI and G's but rather the lifetime of the charge-separated state determines the efficiency of the DNA damage. In other words, the yield of the DNA damage increases with the increase in the lifetime of the charge-separated state.

The combination of the transient absorption measurement and DNA damage quantification provides the data for the effects of the hole transfer on DNA damage during the photosensitized one-electron oxidation of DNA. For the Sens, which can only oxidize G, the close distance between the Sens and G is crucial for efficient DNA damage [4, 58]. In contrast, in the case of Sens, which can also oxidize A to promote hole transfer by A-hopping, the shorter distance between Sens and G is not necessary for efficient DNA damage. This is also true in the case of Schuster's work, where anthraquinone, which can also oxidize A, was used as the Sens [59, 60]. Our results strongly suggest that the hole transfer plays an important role in separating $Sens^{\bullet-}$ and $G^{\bullet+}$ during the photosensitized one-electron oxidation of DNA, providing a sufficiently long time for $G^{\bullet+}$ and $Sens^{\bullet-}$ to react with water or O_2 and avoiding the charge recombination and making the reaction irreversible [61].

5.5 Conclusions

In the present studies, the kinetics of hole transfer in DNA especially by A-hopping was investigated by pulse radiolysis, laser flash photolysis, and their combined method. It was demonstrated that a hole generated on A rapidly migrates through DNA by hopping between A's with a rate constant faster than $10^8\ s^{-1}$ over the distance range of ~30 Å. The rate constant of each A-hopping process between adjacent A's was determined to be ~$10^{10}\ s^{-1}$ from an analysis of the yield of the charge separation depending on the number of A-hopping steps. This fast and weak distance-dependent hole transfer by A-hopping was shown to generate a long-lived, charge-separated state in DNA. When photoinduced electron transfer occurs between excited Sens and A, a part of a hole escapes from the initial charge recombination via consecutive A-hopping to be trapped at G. Once $G^{\bullet+}$ is formed far from Sens$^{•-}$, charge recombination proceeds by either superexchange or A-hopping following slow A oxidation by $G^{•+}$, producing a long-lived, charge-separated state. These results suggest that hole transfer by A-hopping may help to separate the hole and Sens$^{•-}$ during the photosensitized one-electron oxidation of DNA, providing a sufficiently long time for $G^{•+}$ and Sens$^{•-}$ to react with water or O_2. The combination of a transient absorption measurement and DNA damage quantification clearly showed that the yield of the DNA damage correlates well with the lifetime of the charge-separated state during the photosensitized one-electron oxidation of DNA. Our results demonstrate that A oxidation in the consecutive A sequences is crucial for photosensitized DNA damage, suggesting that consecutive A sequences may serve as a good target in photosensitized DNA damage in photodynamic therapy, or that G's adjacent to such sequences may be potential hot spots of oxidative DNA damage.

Acknowledgments

We are deeply indebted to Mr. Tadao Takada for his contributions to these studies. We thank the members of the Radiation Laboratory of ISIR (SANKEN), Osaka University, for running the linear accelerator. This work has been partly supported by a Grant-in-Aid for Scientific Research on Priority Area (417), 21st COE Research, and by others from the Ministry of Education, Culture, Sports and Science and Technology (MEXT) of the Japanese government.

References

1 E. Meggers, M. E. Michel-Beyerle, B. Giese, *J. Am. Chem. Soc.* **1998**, *120*, 12950.
2 P. T. Henderson, D. Jones, G. Hampikian, Y. Z. Kan, G. B. Schuster, *Proc. Natl. Acad. Sci. USA* **1999**, *96*, 8353.
3 M. E. Nunez, D. B. Hall, J. K. Barton, *Chem. Biol.* **1999**, *6*, 85.
4 K. Nakatani, C. Dohno, I. Saito, *J. Am. Chem. Soc.* **1999**, *121*, 10854.
5 F. D. Lewis, T. F. Wu, Y. F. Zhang, R. L. Letsinger, S. R. Greenfield, M. R. Wasielewski, *Science* **1997**, *277*, 673.
6 C. Z. Wan, T. Fiebig, O. Schiemann, J. K. Barton, A. H. Zewail, *Proc. Natl. Acad. Sci. USA* **2000**, *97*, 14052.
7 K. Kawai, T. Takada, S. Tojo, N. Ichinose, T. Majima, *J. Am. Chem. Soc.* **2001**, *123*, 12688.
8 B. Giese, J. Amaudrut, A. K. Kohler, M. Spormann, S. Wessely, *Nature* **2001**, *412*, 318.
9 F. D. Lewis, X. Y. Liu, J. Q. Liu, S. E. Miller, R. T. Hayes, M. R. Wasielewski, *Nature* **2000**, *406*, 51.
10 T. Kendrick, B. Giese, *Chem. Commun.* **2002**, 2016.
11 K. Kawai, K. Miyamoto, S. Tojo, T. Majima, *J. Am. Chem. Soc.* **2003**, *125*, 912.

12 K. Kawai, T. Takada, S. Tojo, T. Majima, *Tetrahedron Lett.* **2002**, *43*, 89.
13 T. Takada, K. Kawai, S. Tojo, T. Majima, *J. Phys. Chem. B* **2003**, *107*, 14052.
14 T. Takada, K. Kawai, S. Tojo, T. Majima, *Tetrahedron Lett.* **2003**, *44*, 3851.
15 S. Steenken, S. V. Jovanovic, *J. Am. Chem. Soc.* **1997**, *119*, 617.
16 C. A. M. Seidel, A. Schulz, M. H. M. Sauer, *J. Phys. Chem.* **1996**, *100*, 5541.
17 I. A. Shkrob, M. C. Sauer, A. D. Liu, R. A. Crowell, A. D. Trifunac, *J. Phys. Chem. B* **1998**, *102*, 4976.
18 V. Shafirovich, J. Cadet, D. Gasparutto, A. Dourandin, W. D. Huang, N. E. Geacintov, *J. Phys. Chem. B* **2001**, *105*, 586.
19 F. D. Lewis, Y. Wu, *J. Photochem. Photobiol. C* **2001**, *2*, 1.
20 J. Jortner, M. Bixon, T. Langenbacher, M. E. Michel-Beyerle, *Proc. Natl. Acad. Sci. USA* **1998**, *95*, 12759.
21 B. Giese, S. Wessely, M. Spormann, U. Lindemann, E. Meggers, M. E. Michel-Beyerle, *Angew. Chem. Int. Ed. Engl.* **1999**, *38*, 996.
22 C. Behrens, L. T. Burgdorf, A. Schwogler, T. Carell, *Angew. Chem., Int. Ed.* **2002**, *41*, 1763.
23 M. Pascaly, J. Yoo, J. K. Barton, *J. Am. Chem. Soc.* **2002**, *124*, 9083.
24 D. M. Guldi, *Chem. Soc. Rev.* **2002**, *31*, 22.
25 H. Imahori, H. Norieda, H. Yamada, Y. Nishimura, I. Yamazaki, Y. Sakata, S. Fukuzumi, *J. Am. Chem. Soc.* **2001**, *123*, 100.
26 D. Gust, T. A. Moore, A. L. Moore, *Acc. Chem. Res.* **2001**, *34*, 40.
27 T. Akiyama, S. Yamada, *Trends Photochem. Photobiol.* **2001**, *8*, 67.
28 D. Porath, A. Bezryadin, S. de Vries, C. Dekker, *Nature* **2000**, *403*, 635.
29 N. C. Seeman, *Angew. Chem. Int. Ed.* **1998**, *37*, 3220.
30 J. J. Storhoff, C. A. Mirkin, *Chem. Rev.* **1999**, *99*, 1849.
31 H. Imahori, K. Tamaki, Y. Araki, Y. Sekiguchi, O. Ito, Y. Sakata, S. Fukuzumi, *J. Am. Chem. Soc.* **2002**, *124*, 5165.
32 F. D. Lewis, R. S. Kalgutkar, Y. S. Wu, X. Y. Liu, J. Q. Liu, R. T. Hayes, S. E. Miller, M. R. Wasielewski, *J. Am. Chem. Soc.* **2000**, *122*, 12346.
33 J. E. Rogers, S. J. Weiss, L. A. Kelly, *J. Am. Chem. Soc.* **2000**, *122*, 427.
34 F. D. Lewis, X. Zuo, J. Liu, R. T. Hayes, M. R. Wasielewski, *J. Am. Chem. Soc.* **2002**, *124*, 4568.
35 T. T. Williams, D. T. Odom, J. K. Barton, *J. Am. Chem. Soc.* **2000**, *122*, 9048.
36 F. D. Lewis, J. Liu, X. Liu, X. Zuo, R. T. Hayes, M. R. Wasielewski, *Angew. Chem. Int. Ed.* **2002**, *41*, 1026.
37 F. D. Lewis, R. L. Letsinger, M. R. Wasielewski, *Acc. Chem. Res.* **2001**, *34*, 159.
38 F. D. Lewis, J. Liu, W. Weigel, W. Rettig, I. V. Kurnikov, D. N. Beratan, *Proc. Natl. Acad. Sci. USA* **2002**, *99*, 12536.
39 F. D. Lewis, Y. Wu, R. T. Hayes, M. R. Wasielewski, *Angew. Chem. Int. Ed.* **2002**, *41*, 3485.
40 M. Bixon, B. Giese, S. Wessely, T. Langenbacher, M. E. Michel-Beyerle, J. Jortner, *Proc. Natl. Acad. Sci. USA* **1999**, *96*, 11713.
41 K. Kino, H. Sugiyama, *Chem. Biol.* **2001**, *8*, 369.
42 S. Shibutani, M. Takeshita, A. P. Grollman, *Nature* **1991**, *349*, 431.

43 D. E. J. G. J. Dolmans, D. Fukumura, R. K. Jain, *Nat. Rev. Cancer* **2003**, *3*, 380.

44 F. D. Lewis, J. Liu, X. Zuo, R. T. Hayes, M. R. Wasielewski, *J. Am. Chem. Soc.* **2003**, *125*, 4850.

45 B. Armitage, *Chem. Rev.* **1998**, *98*, 1171.

46 K. Kawai, T. Takada, S. Tojo, T. Majima, *J. Am. Chem. Soc.* **2003**, *125*, 6842.

47 T. Takada, K. Kawai, X. Cai, A. Sugimoto, M. Fujitsuka, T. Majima, *J. Am. Chem. Soc.* **2004**, *126*, 1125.

48 J. E. Rogers, L. A. Kelly, *J. Am. Chem. Soc.* **1999**, *121*, 3854.

49 D. A. Vicic, D. T. Odom, M. E. Nunez, D. A. Gianolio, L. W. McLaughlin, J. K. Barton, *J. Am. Chem. Soc.* **2000**, *122*, 8603.

50 N. Rahe, C. Rinn, T. Carell, *Chem. Commun.* **2003**, 2120.

51 K. Kobayashi, S. Tagawa, *J. Am. Chem. Soc.* **2003**, *125*, 10213.

52 B. Giese, A. Biland, *Chem. Commun.* **2002**, 667.

53 B. Giese, M. Spichty, *ChemPhysChem* **2000**, *1*, 195.

54 C. Dohno, A. Ogawa, K. Nakatani, I. Saito, *J. Am. Chem. Soc.* **2003**, *125*, 10154.

55 F. D. Lewis, X. Liu, S. E. Miller, R. T. Hayes, M. R. Wasielewski, *J. Am. Chem. Soc.* **2002**, *124*, 14020.

56 C. Dohno, E. D. A. Stemp, J. K. Barton, *J. Am. Chem. Soc.* **2003**, *125*, 9586.

57 J. Yoo, S. Delaney, E. D. A. Stemp, J. K. Barton, *J. Am. Chem. Soc.* **2003**, *125*, 6640.

58 B. Giese, *Acc. Chem. Res.* **2000**, *33*, 631.

59 L. Sanii, G. B. Schuster, *J. Am. Chem. Soc.* **2000**, *122*, 11545.

60 G. B. Schuster, *Acc. Chem. Res.* **2000**, *33*, 253.

61 K. Kawai, T. Takada, T. Nagai, X. Cai, A. Sugimoto, M. Fujitsuka, T. Majima, *J. Am. Chem. Soc.* **2003**, *125*, 16198.

6
Chemical Probing of Reductive Electron Transfer in DNA

Steven E. Rokita and Takeo Ito

6.1
Introduction

Research on charge transfer in DNA is propelled by a diverse set of interests broadly ranging from mutagenesis to nanoelectronics. The influence of charge transfer on the formation of DNA lesions is evident from its potential to uncouple the initial sites of redox reaction from the ultimate sites of covalent products [1–4]. Conductivity of DNA serves a more benign function when applied to sensors and electronic devices [5–10]. These latter functions also rely on the exquisite ability of DNA to assemble into predictable and complex structures [11] that can be responsive to analytes [12]. Related advances and innovations have been possible through the collective efforts of many experimentalists and theorists who explore fundamental variables affecting transfer processes. Current and future success in these areas depends greatly on our thorough and accurate understanding of charge formation, maintenance, and diffusion in DNA.

Many of the principles controlling charge transfer in DNA have been characterized through photoinduced single-electron transfer (PSET). Typically, chromophores are covalently tethered to defined sequences of DNA and charge separation is initiated by ultraviolet irradiation. The extent of transfer depends on competition between charge recombination, further separation, and quenching of various types. If the excited-state chromophore acts as an oxidant, then a center of electron deficiency is formed in DNA that is able to migrate in a process often described as hole transfer (HT) (Scheme 6.1). If the excited-state chromophore acts instead as a reductant, then a center of electron excess is formed in DNA that may migrate in an complementary process often described as excess electron transfer (EET) (Scheme 6.1). These processes both depend on transfer of an electron. However, the primary charge carriers and molecular orbitals involved in transfer differ. Our understanding of HT has benefited from considerable attention over the last decade and is now well advanced. Theoretical and experimental treatment of EET began only quite recently, and even its most basic features still need to be identified [13–19].

Charge Transfer in DNA: From Mechanism to Application. Edited by H.-A. Wagenknecht

ISBN: 3-527-31085-1

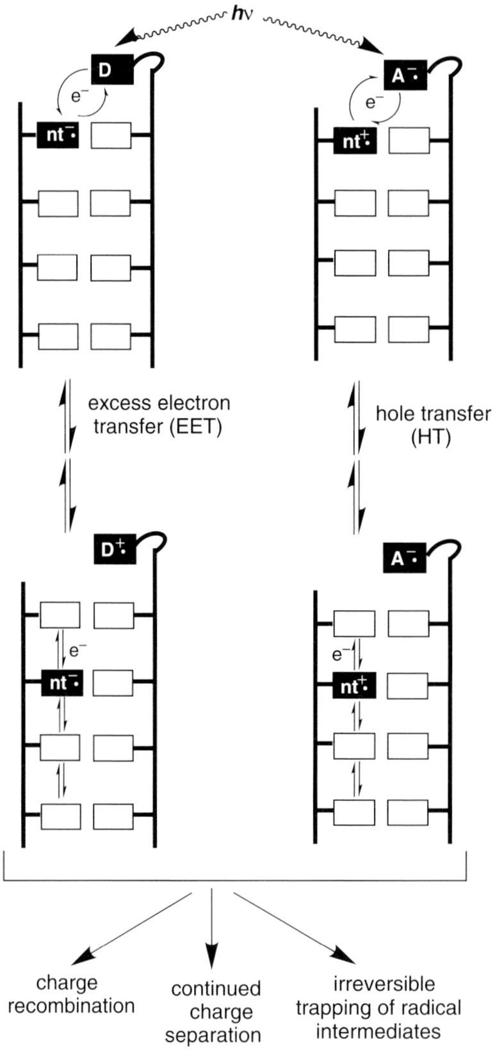

Scheme 6.1 Photoinduced single-electron transfer may induce excess electron or hole transfer in DNA.

Spectroscopic and chemical probes of HTand EETallow observation over time regimes of picoseconds to minutes, and each have unique advantages and limitations. Electron paramagnetic resonance [20, 21] and transient absorbance [22–25] spectroscopies provide insight into the nature and kinetics of reactive intermediates but offer little information on irreversible transformations induced by charge transfer. In contrast, detection of covalent changes subsequent to transfer focuses attention on net transformations but provides little direct information on the intermediate species formed by competitive pathways. Chemical methods also generate explicit information on the distance that can be traversed in DNA between product generation and electron abstraction or donation. This is particularly relevant when studying the distribution and sequence dependence of particular lesions [26–28] or developing new DNA-based materials with enhanced transfer capabilities [29].

Each experimental system created for studying charge transfer in DNA is framed by a choice of methods for injecting charge into DNA and observing its ensuing migration. Results from individual systems may often appear to vary widely, but they do not necessarily reflect contradictions in the nature of charge transfer. Instead, these differences can originate from unforeseen properties that are unique to a specific injection, detection, or DNA system. The principles underlying charge transfer emerge only after comparing data from numerous approaches and systems. This chapter reviews use of BrdU (^{Br}U) for detecting EET and aromatic amines for injecting an electron into DNA. Observations from our laboratory based on this experimental design are also summarized and compared with the limited number of other chemical approaches used to date for exploring EET. Although the aromatic amines have been applied only in conjunction with ^{Br}U reduction, there is no inherent advantage of this pairing. Electron injection by a ketyl radical [30] or a reduced flavin [31] would likely function equally well to reduce ^{Br}U. Similarly, all three donors should also support other methods of trapping excess electrons as well as spectroscopic detection of transient intermediates.

6.2 A Reaction-based Method to Characterize Excess Electron Transfer

6.2.1 A Chemical Probe for Excess Electron Transfer

The effect of ^{Br}U substitution in DNA has been studied for over 40 years [32, 33]. This residue is nearly isosteric to thymidine and typically pairs with adenosine when incorporated into DNA, although occasional mispairing with guanine likely contributes to its mutagenic effect [34, 35]. Investigators quickly realized that DNA containing ^{Br}U was also highly sensitive to degradation by both ultraviolet light and radiolysis [32, 33, 36]. Such treatment of DNA containing ^{Br}U caused strand scission, base release, and cross-linking. This discovery stimulated many additional studies on the photochemistry of ^{Br}U and related model compounds, and similar efforts then followed on the other halouridines, 5-fluoro-, 5-chloro-, and 5-iodouridine. A consensus about the photolability of ^{Br}U ultimately converged around its ability to accept an electron, eliminate bromide, and generate an intermediate uridinyl radical. The fate of this radical depends on specific reaction conditions and the local structural and solvent environment [37–40].

Of all the 5-halouridine derivatives, ^{Br}U was most appealing for study of EET due to its optimum balance of physical and chemical properties. This nucleobase has already received most of the attention devoted to the photochemistry of the halogenated series. In addition, its phosphoramidite is commercially available, along with the iodo and fluoro derivatives, for oligodeoxynucleotide synthesis by standard automated solid-phase methods. All of the halogenated pyrimidines are ex-

could also be performed [41]. Concurrently, the Carell laboratory developed a flavin nucleotide for incorporation into DNA that is capable of injecting an electron into DNA when irradiated in its reduced form at 360 nm [31, 57]. This latter system was inspired by DNA photolyase, a natural enzyme that utilizes an excited-state reduced flavin to repair thymine dimers in DNA [58]. Aromatic amines are also known to promote monomerization of thymine dimer under UV irradiation [59], and this class of compounds offers an attractive alternative for electron injection into DNA. A broad range of aromatic amines are commercially available or easily synthesized, and a large subset of these have the potential to satisfy the above criteria.

An expedient assay was developed to identify aromatic amines that might serve best as electron donors for EET without the need to generate all of the corresponding oligodeoxynucleotide conjugates [60]. Potential donors were screened with duplex DNA containing an abasic site to facilitate their binding and stacking and a ^{Br}U residue to provide an irreversible trap for EET (Scheme 6.3). One of the simplest representatives, dimethylaniline, was not tested in this assay despite its ability to promote PSET and thymine dimer cleavage in model systems [59]. The absorption maxima of this and related derivatives are not sufficiently shifted away from the absorption of DNA to allow for selective excitation. Additionally, these derivatives were not expected to associate strongly with duplex DNA. In contrast, aminoanthracene derivatives would likely stack well within an abasic site [61, 62]. However, these alternative aromatic amines are not strong enough electron donors in their excited state to inject an electron into duplex DNA [60].

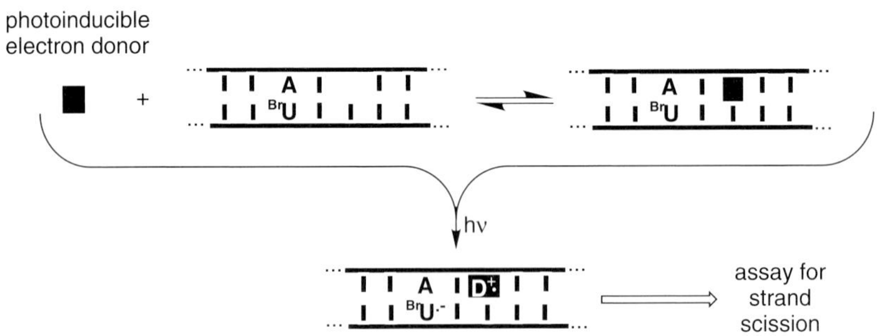

Scheme 6.3 Activity-based screen for compounds that bind to an abasic site and induce EET in DNA.

The first compounds tested in our screen were *N,N,N′N′*-tetramethylbenzidine (TMB), *N,N,N′N′*-tetramethyl-1,5-diaminonaphthalene (TMDN), and 1,5-diaminonaphthalene (DAN). Each absorbed light above 335 nm to avoid direct irradiation of ^{Br}U (Figure 6.1). Also, the excited-state potential for each was sufficient to reduce all of the natural nucleotides [60]. DNA strand scission based on ^{Br}U reduction was evident after a 10-min irradiation ($\lambda > 335$ nm) in the presence of 0.5 mM TMDN (Figure 6.2). DAN exhibited greater efficiency and induced scission under the same conditions using concentrations as low as 0.1 mM. No activ-

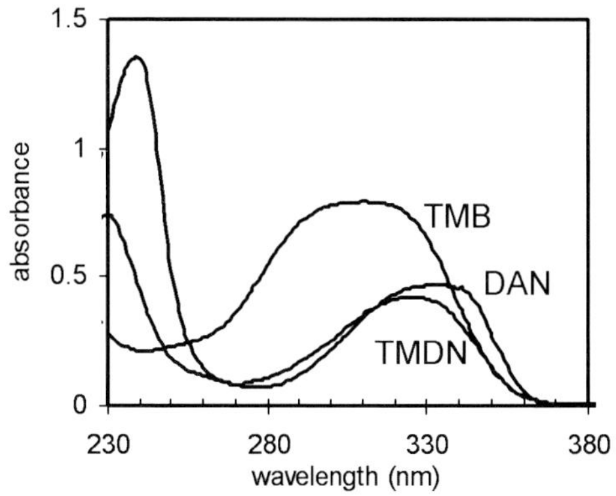

Fig. 6.1 UV absorption spectra of TMDN, TMB, and DAN (50 μM) in acetonitrile.

ity was observed with TMB even at concentrations of 1.0 mM. Lack of efficient electron injection by TMB is likely due to its inability to stack into helical DNA. Perhaps this chromophore might gain the ability to stack properly after covalent conjugation to the oligodeoxynucleotide containing the abasic site, but our screen was designed to identify a dual capacity of donating an electron and binding intrinsically to an abasic site. For example, PSET and strand scission by TMDN required the abasic site, and the absence of this site prevented ^{Br}U reduction in duplex DNA. Similarly, ^{Br}U was crucial for inducing strand scission, and DNA lacking this residue, but retaining an abasic site, was also unreactive to TMDN [63].

Oligodeoxynucleotide conjugates of TMDN were next prepared to begin characterizing the sequence and distance dependence of EET as revealed by electron injection from an aromatic amine and electron addition to ^{Br}U. Covalent attachment constrains the location of the donor and avoids the need for addition of excessive concentrations of a free donor such as TMDN. Comparing the results of the TMDN-based system with those of other systems designed for EET should help to distinguish which properties are unique to each set of donors and acceptors and which are fundamental to all EET in DNA. In an equivalent manner, future com-

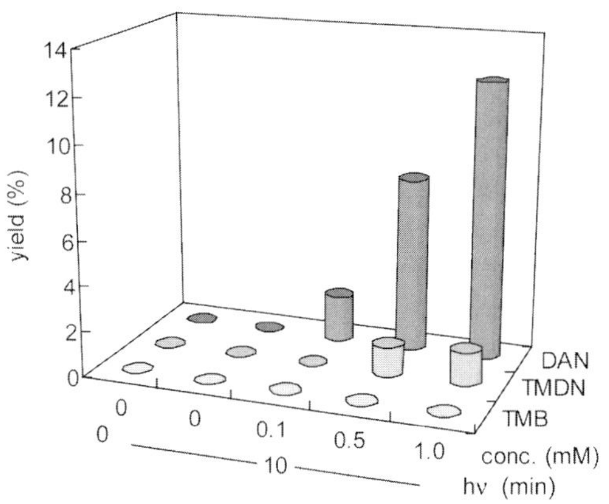

Fig. 6.2 Strand scission (percentage) induced by EET from non-conjugated aromatic amines to a duplex containing an abasic site and a ^{Br}U residue [60]. Samples were photoirradiated (λ > 335 nm) in the alternative presence of TMB, TMDN, and DAN under anaerobic conditions and then treated with hot piperidine and analyzed by gel electrophoresis.

parisons between oligodeoxynucleotide conjugates of TMDN, DAN, and other active aromatic amines will provide crucial information on the classic topic of structure vs. activity within a closely related system. Despite the greater activity of DAN vs. TMDN, investigation of DAN conjugations was deferred due to possible complications created by reported side reactions with DNA [64].

The abasic site that was integral to the activity screen above was also useful as a site for attaching suitably prepared electron donors. An enormous variety of linking strategies are available and many have been used in studies on HT and EET. The abasic site allowed for conjugation of the donor after automated DNA synthesis, thus circumventing the need to synthesize a phosphoramidite analogue of the donor. The particular derivative used to form the abasic site is commercially available as a phosphoramidite with the otherwise reactive aldehyde protected as a diol. Oxidation of the diol with periodate generates the aldehyde when desired [65]. Such aldehydes typically condense with amines and oxyamines quite readily. For TMDN, oxyamine coupling has been most reliable (Scheme 6.4). Placement of the abasic site in a central location within duplex DNA supported comparisons of EET through intra- vs. interstrand pathways in a 3′ to 5′ and 5′ to 3′ direction (see below) [66].

An optimal linker between an oligodeoxynucleotide and its chromophores used for HT or EET will promote favorable stacking within helical DNA, and this in turn promotes optimum charge injection. Systematic analysis of linker length and composition had previously been pursued by laboratories interested in antisense activities of DNA conjugates, and a few equivalent studies have been published on charge transfer [67]. Most often, linkers are selected from qualitative guidelines. Short or rigid linkers are avoided unless they are very carefully designed for fear of destabilizing possible intercalation. Similarly, very long and flexible linkers are avoided for fear of establishing a broad diversity of conformations. A convenient and appropriate linker for TMDN was not initially obvious, and thus two derivatives were prepared for investigation. ODN **1**$'_4$ was constructed with a $-(CH_2)_4-$ linker connecting the oxyamine and aromatic nitrogen of TMDN, and ODN **1**$'_7$ was constructed with an equivalent $-(CH_2)_7-$ linker (Scheme 6.4).

DNA duplexes formed by ODN **1**$'_4$, ODN **1**$'_7$, and a complementary oligodeoxynucleotide exhibited equivalent melting temperatures (53–54 °C). These values are

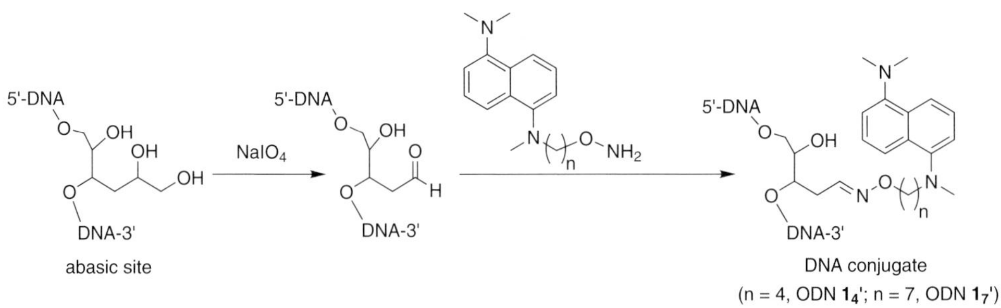

Scheme 6.4 Covalent coupling of an electron donor to DNA.

Y =
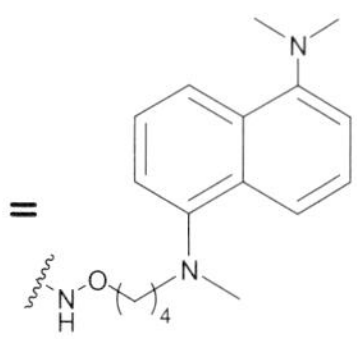

ODN $\mathbf{1}_{4/7}'$	5′ CGT CAT G**Y**A ATG TAC TGC 3′	
ODN **2**	3′ GCA GTA C**G**T ^{Br}**U**AC ATG ACG 5′	
ODN **3**	GCA GTA C**C**T ^{Br}**U**AC ATG ACG	
ODN **4**	GCA GTA C**T**T ^{Br}**U**AC ATG ACG	
ODN **5**	GCA GTA C**A**T ^{Br}**U**AC ATG ACG	
ODN **6**	5′ GCC GG^{Br}**U** TTG TTT GGC GC 3′	
ODN **6′**	3′ CGG CCA AA**Y** AAA CCG CG 5′	
ODN **7**	5′ GCC GGT TTG AAA GGC GC 3′	
ODN **7′**	3′ CGG CCA AA**Y** TT^{Br}**U** CCG CG 5′	
ODN **8**	5′ GCC GGT TTG TT^{Br}**U** GGC GC 3′	
ODN **9**	5′ GCC GGA AAG TTT GGC GC 3′	
ODN **9′**	3′ CGG CC^{Br}**U** TT**Y** AAA CCG CG 5′	

Chart 6.1 Oligodeoxynucleotide sequences used to study EET.

somewhat lower than that obtained for a comparable sequence of native DNA (63 °C) but higher than that obtained for the corresponding duplex containing the abasic site (50 °C) [60, 63]. ODN $\mathbf{1}_4'$ was approximately 3.4-fold more efficient at EET than ODN $\mathbf{1}_7'$ when ODN **2** containing ^{Br}U was used as the complementary DNA strand and irradiation (λ > 335 nm) was sustained for 5 min (Chart 6.1 and Figure 6.3) [60]. Over longer periods of irradiation, their difference in activity diminished to less than twofold, suggesting that the TMDN bound to ODN $\mathbf{1}_7'$/ODN **2** might stack less efficiently or orient in a variety of modes compared to the TMDN bound to ODN $\mathbf{1}_4'$/ODN **2**. The fluorescence emission of the two TMDN derivatives provided the first evidence of their different environments within duplex DNA (Figure 6.4). The chromophore of ODN $\mathbf{1}_7'$/ODN **2** was less susceptible to quenching ($\Phi_F = 0.018$) and its emission maximum was red-shifted ($\lambda_{max} = 454$ nm) relative to that observed for ODN $\mathbf{1}_4'$/ODN **2** ($\Phi_F = 0.015$, $\lambda_{max} = 444$ nm) [60].

The photoexcited-state properties of the oligodeoxynucleotide-bound TMDN derivative and its ability to inject electrons into DNA were also expected to vary depending on the neighboring nucleotides. Already, adjacent nucleotides on the same strand as attached pyrene and fluorene derivatives have been shown to influence the yield of emission [24, 68], and comparable results are anticipated for TMDN conjugates. Nucleobases proximal to the chromophore, but on the comple-

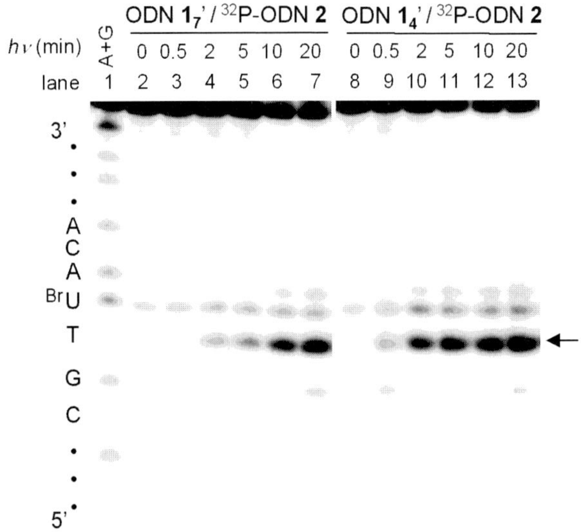

Fig. 6.3 Phosphoimage of a 20% denaturing polyacrylamide gel showing strand scission induced by EET. 5-[^{32}P]-labeled ODN **2** was alternatively annealed with ODN $\mathbf{1}_7'$ and ODN $\mathbf{1}_4'$, irradiated for the indicated time, and treated with hot piperidine.

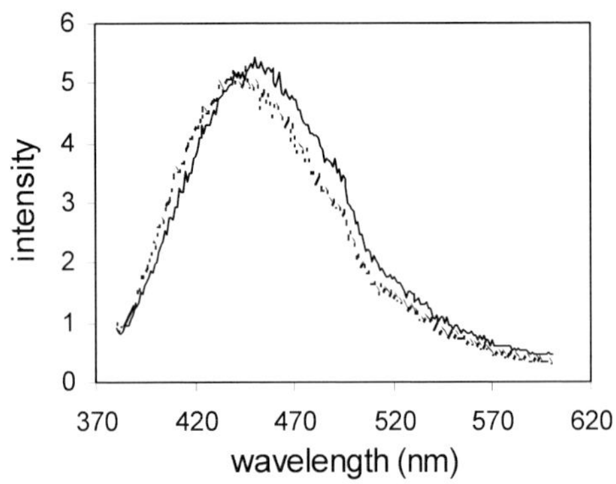

Fig. 6.4 Fluorescence spectra of ODN $\mathbf{1}_4'$/ODN **2** (dotted line) and ODN $\mathbf{1}_7'$/ODN **2** (solid line) in 10 mM sodium phosphate (pH 7.0) and 90 mM NaCl upon excitation at 330 nm.

mentary strand, also have the potential to influence its photochemical properties. This was particularly likely for the nucleobase directly counter to the TMDN-containing residue.

A series of oligodeoxynucleotides complementary to ODN $\mathbf{1}_4'$ were prepared to examine the influence of G (ODN **2**), C (ODN **3**), T (ODN **4**), and A (ODN **5**) as counter bases (Chart 6.1). The relative yields of electron injection and subsequent DNA cleavage based on ^{Br}U reduction were indeed greatly dependent on the counter base. The nucleobases C and T (–2.59 V and –2.42 V vs. SCE, respectively [69]) supported the least cleavage of DNA after irradiation (Figure 6.5 a). In contrast, these same sequences were most effective at quenching fluorescence of the at-

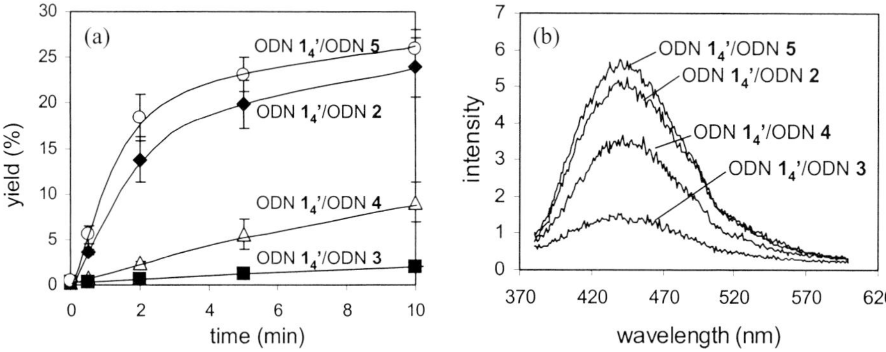

Fig. 6.5 (a) Efficiency of strand scission induced by EET depends on the nature of the nucleobase counter to the electron donor. ODN $1_4'$ was annealed with the complementary strands (ODN **2–5**) and irradiated (>335 nm) for the indicated periods. Samples were then treated with hot piperidine and analyzed by PAGE [60]. (b) Fluorescence spectra of duplexes formed by ODN $1_4'$ and ODNs **2–5** alternatively upon excitation at 330 nm [60].

tached TMDN derivative (E^*_{ox} = 2.80 V [63]) (Figure 6.5 b). The greatest yield of cleavage was induced by G and A counter bases (–2.76 V and <–3.00 V vs. SCE, respectively [69]), which were conversely the least effective at quenching the attached TMDN derivative. These results are rather counterintuitive if fluorescence quenching, electron injection, and EET are equated to one another. Certainly, both EET and fluorescence quenching depend on the efficiency of electron injection into DNA, but only EET is also affected by competing processes such as back electron transfer. Thus, EET may be maximized by tethering an electron donor for optimum stacking within helical DNA at a nucleobase sequence that promotes electron injection and suppresses back electron donation and other processes that would quench the excess electron.

6.3 Structural Dependence of ^{Br}U Reduction in DNA by Excess Electron Transfer

6.3.1 Electrons Donated by an Aromatic Amine Remain Associated with the Nucleobases of DNA

Modulation of EET by nucleobase sequence and distance was apparent from the very earliest studies of DNA containing an aromatic amine donor and a ^{Br}U acceptor [63]. However, explicit correlations between DNA structure and EET efficiency required confirmation that the nucleobases actually participated directly in electron transfer. Just as TMDN could inject an electron into DNA, this donor might also release an electron into solution. A solvated electron in turn might then diffuse and recombine with DNA or, more specifically, ^{Br}U. This alternative path is typically identified by its quenching in the presence of N_2O. No such quenching was detected in oligodeoxynucleotide systems containing a TMDN derivative; consequently, excess electron density appears to remain associated with the DNA duplex [63].

EET from a TMDN conjugate to ^{Br}U is also not very sensitive to the presence of molecular oxygen. Most EET and many HT experiments are commonly performed under anaerobic conditions to avoid quenching by molecular oxygen. Aerobic conditions suppressed EET in our system by a maximum of approximately 30% relative to anaerobic conditions. The origins of this effect have not yet been determined but likely originate from competing electron transfer to molecular oxygen. Alternative addition of molecular oxygen to nucleobase radicals would cause strand scission at a variety of sites in DNA subsequent to piperidine treatment, and no such patterns of scission were detected under aerobic conditions [60]. As described above, molecular oxygen is also not expected to interfere with the ^{Br}U-dependent assay for EET [37, 39, 53].

Photoinduced strand scission of ODN $\mathbf{1}_4'$/ODN **2** was inhibited by a high concentration of 2-mercaptoethanol (ME, 50 mM) although little effect was observed at low concentrations (5 mM) (Figure 6.6). This thiol has the potential to alternatively reduce and couple to the intermediate uridinyl radical and thus protect against strand scission [51]. Reduction of the radical by ME to form the dehalogenated uridine derivative seems improbable since another strong reductant (dithionite, 50 mM) did not significantly inhibit strand scission. Thus, the intermediate radical is likely intercepted by addition of the thiol. This radical can also be expected to show similar sensitivity to other reagents that couple equivalently. Studies are currently directed at determining the products formed by irradiation of the oligodeoxynucleotides above in the presence and absence of thiols.

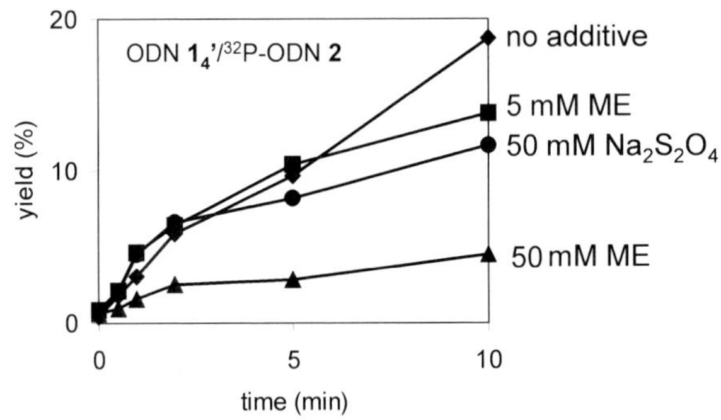

Fig. 6.6 Effect of added reductant on the yields of strand scission after irradiation and piperidine treatment under standard conditions. Samples were exposed to UV light ($\lambda > 335$ nm) in the alternate presence or absence of 5 mM and 50 mM 2-mercaptoethanol (ME) or 50 mM sodium dithionite [60].

6.3.2
Distance, Sequence, and Directional Dependence of EET in Duplex DNA

The great distances traversed by HT in DNA were among the first characteristics noted for this phenomenon [70]. This feat is now generally believed to result from a multi-step hopping mechanism that relies primarily on guanine residues for carrying the electron deficiency. A comparable mechanism can be envisioned for EET, but in this case, pyrimidine residues act as carriers of the electron excess. These nucleobases are most easily reduced and known to accept electrons from PSET and pulse radiolysis [21, 69, 71, 72]. The first study of EET in DNA involving a systematic increase in the distance between a donor and acceptor relied on the ability of excess electron density to promote reversion of a thymine dimer [31]. In this example, 1–7 contiguous A-T base pairs were used to separate the donor and acceptor and a weak logarithmic distance dependence (slope = 0.1 $Å^{-1}$) was detected [73]. A slightly greater dependence on distance (slope = 0.3 $Å^{-1}$) was detected in our initial characterization of EET using a TMDN derivative connected through a $-(CH_2)_4-$ bridge to an abasic site that was positioned zero to five base pairs from a ^{Br}U [63]. The moderate increase in distance dependence was originally ascribed to the presence of intervening G-C pairs, but more likely the correlation itself was merely fortuitous. Distance has recently emerged as just one of many variables contributing to the efficiency of EET, as described below.

The chemical processes that limit EET in DNA are only beginning to be identified. Initial electron injection apparently competes with charge recombination (back electron transfer) in both EET and HT [25, 56, 60, 74–76]. Once a diffusible charge is out of range of efficient recombination, other mechanisms may begin to attenuate successive transfer steps. Molecular oxygen is not especially effective at trapping excess electron density, as is evident from its minor perturbation of long-distance reduction of ^{Br}U [60]. Currently, the most important determinant of EET appears to be the specific distribution of thymine residues between the donor and acceptor [25]. Cytosine residues may accept electrons with nearly the same propensity as thymine residues, but the resulting cytosinyl radical is subject to rapid protonation from its paired guanine. This neutralizes and hence stabilizes the radical. Further transfer becomes energetically less favorable. Radiolysis of polynucleotide sequences indicates that such proton transfer inhibits but does not terminate EET [77]. Participation of C during EET was in part demonstrated in this study by a solvent isotope effect from substituting D_2O for H_2O. Our system also revealed a loss of EET efficiency when two A-T base pairs between the electron donor and acceptor were replaced by two G-C pairs [66]. Similarly, substituting D_2O for H_2O enhanced EET through the G-C pair by replacing proton transfer with slower deuteron transfer. No comparable solvent isotope effect was observed for EET through A-T base pairs.

Investigations are only now beginning to explore the detailed response of EET to systematic perturbations in nucleotide sequence and donor-acceptor juxtaposition. However, two basic principles have been suggested by our early studies based on an aromatic amine donor and a ^{Br}U acceptor. First, transfer or hopping of an

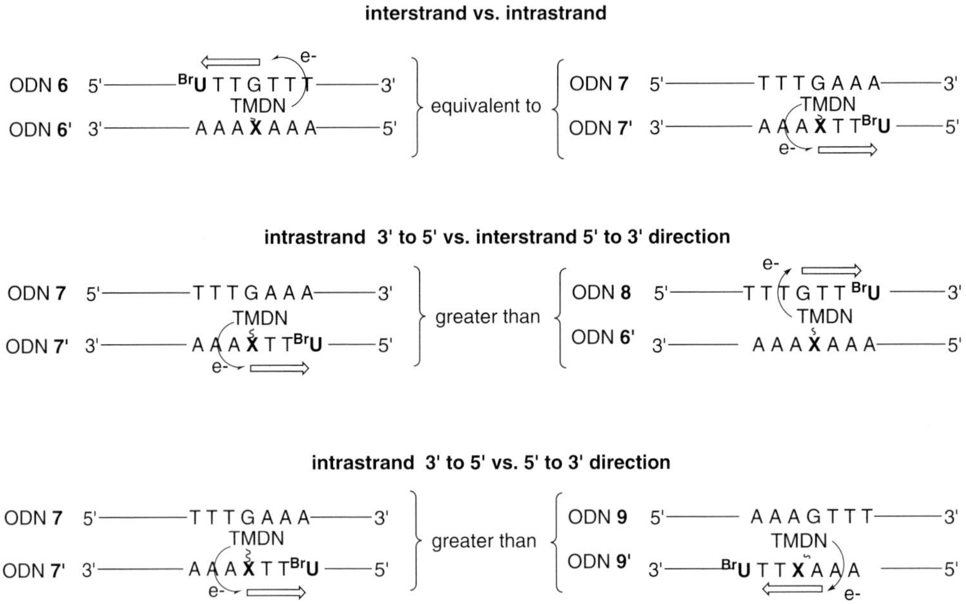

Scheme 6.5 Efficiency of intra- vs. interstrand EET from the 3′ to 5′ and 5′ to 3′ directions.

excess electron is not inhibited by enforcing an interstrand rather than intrastrand transfer [66]. The efficiency of EET from a 3′ to 5′ direction in ODN **6**/ODN **6’** and ODN **7**/ODN **7’** differed by less than 10% (Scheme 6.5). In contrast, converting intrastrand transfer with a 3′ to 5′ directionality (ODN **7**/ODN **7’**) to interstrand transfer with a 5′ to 3′ directionality (ODN **8**/ODN **6’**) decreased the efficiency of EET by more than 75%. Second, the importance of directionality was verified by comparing two systems that both support intrastrand transfer from the donor to acceptor. Transfer in the 5′ to 3′ direction as represented by ODN **9**/ODN **9’** was less than 2% of that from 3′ to 5′ as represented by ODN **7**/ODN **7’** [66]. EET is also suppressed when the route of transfer via thymine residues alternates between strands multiple times.

Further study is still necessary to verify the initial structural correlations as well as to discover other major and minor determinants controlling EET in DNA. In particular, the influence of the nucleotides that neighbor the TMDN and ^{Br}U residues requires attention. Those surrounding TMDN could affect its alignment within helical DNA and its photochemical properties such as absorptivity and excited-state relaxation. Those surrounding ^{Br}U could affect its reduction and subsequent hydrogen atom abstraction from its 5′ neighbor. The net yield of hydrogen abstraction is not expected to express a significant conformational dependence since strand scission by ^{Br}U reduction appeared comparable in single- and double-stranded DNA [39, 53]. However, theoretical studies suggest that the electron affinity of ^{Br}U may vary depending on its adjacent nucleotides [43]. This in turn would affect the driving force for trapping the excess electron. At least the oligodeoxynucleotides used in the specific comparison above all contained the same counter base. Thus, this important variable did not obscure the features described above that also affect EET in DNA.

6.3.3
Significance of Indirect Assays for Detecting EET

The characteristics of EET in DNA detailed above are most accurately described when considered together with the detection method applied for analysis, which, in this case, is based on ^{Br}U reduction and an ensuing strand scission. Neither chemical nor physical assays truly observe EET. Rather, the consequences of EET are measured. Spectroscopy can follow the formation and consumption of fleeting intermediates, while chemical methods focus on those intermediates that undergo irreversible covalent modification. Both approaches offer information on competition between electron migration and quenching. Similarly, studies on HT detect competition between hole migration and trapping by molecular oxygen or water [45–47].

Only two chemical assays have so far been used to detect EET in DNA. The first, monomerization of thymine dimer, has the advantage of not consuming the excess electron. The extent of electron diffusion can hence be traced through multiple chemical events [30, 31]. The rate of dimer monomerization has been estimated to occur on the order of a microsecond [59], which in turn provides little information on picosecond and nanosecond processes of EET such as electron injection and initial charge diffusion. In contrast, the assay based on ^{Br}U quenches and consumes the excess electron but has the potential to report on a time regime of nanoseconds as set by the estimated rate of Br^- elimination from $^{Br}U^{\bullet -}$ [44]. The differing rates of these two trapping methods have recently been proposed as the cause for conflicting data on the sensitivity of EET to nucleotide sequence and transfer direction [30]. Monomerization of thymine dimer revealed no dependence on DNA sequence between the acceptor dimer and the donor flavin [78, 79], unlike the results based on an acceptor ^{Br}U and a donor amine [66]. If the rates of trapping EET are indeed central to these conflicts, then use of an analogous ^{Cl}U in the place of ^{Br}U should yield data similar to that generated by dimer monomerization. Consumption of the intermediate $^{Cl}U^{\bullet -}$ and the dimer radical anion occurs over the microsecond rather than the nanosecond range [44]. Ultimately, EET will best be described after numerous chemical and physical approaches are combined together to span the full range of time regimes. Concurrent use of diverse electron donors will similarly help to establish a broad understanding of this interesting process.

7.2
Hole-injection Systems

Development and evaluation of hole-injection systems, which can generate the base radical cation (hole) selectively at a desired position anywhere in the DNA, are a very important part of HT studies. We have devised a hole-injecting nucleoside, cyanobenzophenone (CNBP)-substituted 2′-deoxyuridine (^{CNBP}U), that contains an electron-accepting CNBP chromophore at the C5 position of 2′-deoxyuridine [16]. CNBP is a good electron-accepting photosensitizer [17] that selectively damages DNA at 5′G of the GG sequence, which is a landmark of one-electron oxidation of DNA [18–20]. Incorporation of CNBP chromophore into DNA enables us to determine the hole-injection site and to discuss a remote oxidation via HT through DNA. Molecular modeling studies indicate that CNBP chromophore is placed in the major groove of the ^{CNBP}U-containing DNA helix, and thus ^{CNBP}U can be introduced into oligodeoxynucleotide (ODN) duplexes at predetermined sites without perturbing the standard B-form structure. We examined a series of photoreactions using a variety of ODNs that contain ^{CNBP}U and GG hole traps [5, 21]. The efficiency of remote G oxidation is highly dependent on the location of the nearest G to ^{CNBP}U, which is responsible for the hole-injection step. The most effective sequence for the hole injection is 5′C^{CNBP}U/AG. When the nearest G base is located in the same strand containing ^{CNBP}U, the remote oxidation is most evident in the sequence containing 5′GT^{CNBP}U/AAC. Schematic illustration of HT initiated by ^{CNBP}U is shown in Figure 7.1. Guanine radical cation ($G^{\bullet +}$, hole) is injected site-selectively at the proximal G by single-electron transfer from G to photoexcited ^{CNBP}U. The hole generated in DNA migrates to the distal GG hole trap and is eventually trapped by O_2 and/or water to give alkali-labile sites. The most effective sequence, 5′C^{CNBP}U/AG, is used for HT studies in the following sections.

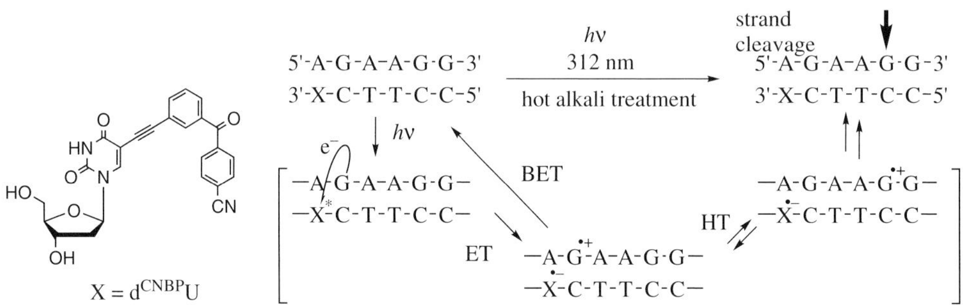

Fig. 7.1 Schematic representation of remote G oxidation by cyanobenzophenone-substituted uridine (X = d^{CNBP}U).

7.3 Modulation of Hole Transport Efficiency

7.3.1 Modulation of Hole Transport Efficiency by Artificial Bases

The DNA double helix contains an extended array of π-stacked aromatic bases that are responsible for the charge conductivity in DNA. Thus, chemical properties of the bases play a crucial role in determining HT efficiency. We have designed and synthesized a number of modified bases for modulation of HT efficiency. Efforts to modulate HT efficiency will provide useful information for understanding the details of the HT mechanism as well as for applying DNA-based nanomaterials. Our first aim is to synthesize artificial nucleosides that increase HT efficiency. HT efficiency significantly varies depending on a number of factors, such as base stacking, ionization potential, dynamic motion, and ionic atmosphere. Since these factors correlate with each other, it is not straightforward to design the artificial bases possessing high HT ability. First, HT should be accelerated by increasing the thermodynamic driving force [22]. Such energetic control is easily achieved by replacement of natural bases with an artificial base that has a lower oxidation potential [8]. Introduction of the electron-donating bases can accelerate the HT rate by directly increasing the driving force between electron donor (artificial base) and acceptor (natural bases). Another way to improve HT ability is the expansion of π-way. Artificial bases possessing expanded aromatic surface will provide better HT ability [9], since the enhanced π-stacking greatly increases base-base electronic interactions.

7.3.1.1 Ionization Potential of Bridged Bases

Ionization potential (IP) of the bases is deeply involved in the energetics of HT through DNA. In order to know the effect of IP of bridged bases on HT efficiency, we examined HT between two GGG sites separated by a TTBTT bridged sequence, where the bridged base (B) is A(ODN **A2**), 7-deazaA (^{Z}A, ODN ^{Z}**A2**), 2,6-diaminopurine (^{da}P, ODN ^{da}**P2**), G (ODN **G1**), or 7-deazaG (^{Z}G, ODN ^{Z}**G1**) (Figures 7.2 and 7.3) [8]. The bridged bases have distinct oxidation potentials with minimal structural distortion of the B-form duplex. Upon photoirradiation of the duplexes, the hole is site-selectively generated at G_2 by a single-electron transfer to ^{CNBP}U opposite A_1 in the complementary strand (ODNs **X1** and **X2**). Since consecutive G sequences have the lowest IP in standard B-form DNA, radical cation is likely localized in a proximal GGG site (GGG_p) at first. Hole is next transported to a distal GGG site (GGG_d) across the TTBTT bridge with different rates depending on the nature of the bridged base B (Figure 7.3). Oxidation potentials of the nucleosides were estimated from the measurements of cyclic voltammograms in water containing 0.1 M $LiClO_4$ (Figure 7.2). The measured oxidation potentials of A, ^{Z}A, ^{da}P, G, and ^{Z}G were 1.39 V, 1.15 V, 0.96 V, 1.07 V, and 0.74 V, respectively. Molecular orbital calculation by DFT/B3LYP/6-31G(d) gave IP values in the same order as the measured oxidation values.

also allow the existence of positive charge on bridged bases other than guanine bases [1, 2]. Although the precise mechanism still remains to be resolved, these results clearly show that HT efficiency between two GGG sites can be sensitively modulated by IP of the bridged base. It should be added that the sequence dependency is not necessarily the same for other oxidants. HT initiated by ^{CNBP}U is highly sequence dependent, but different oxidants can alter the sequence dependency due to the different HT kinetics and energetics [7].

7.3.1.2 Enhanced Base Stacking

Creating an artificial DNA that functions as a much better HT mediator is an intriguing subject. In the previous section, we showed that the rate of HT was increased by using modified nucleosides possessing lower IP. However, this strategy is not effective for improvement of inherent conductivity of DNA. In addition, lowering IP makes the modified base more susceptible to hole-trapping reactions. Our group introduced an alternative concept of rational design for creating a much better mediator for HT. Benzodeazaadenine (^{BD}A) derivatives have the expanded aromatic system, which increases HT efficiency owing to the enhancement of π-stacking [9]. In general, HT efficiency falls off with increasing distance from a hole acceptor. The bridged sequence consisting of the consecutive ^{BD}A's did not decrease the HT efficiency between two GGG hole traps; rather, it increased with increasing the number of ^{BD}A runs up to five ^{BD}A units (Figure 7.4 b). It has been proposed that stationary polarons in DNA can extend over 5–7 base pairs, and hence the influence of the length of the ^{BD}A runs on the HT efficiency would possibly relate to the polaron width generated in DNA [2, 26]. Figure 7.4 c shows the sequencing gel analysis of HT through 20 consecutive methoxy-substituted ^{BD}A (^{MD}A, R = MeO) units. The I_{Gd}/I_{Gp} value was 0.99±0.04, implying that a hole in the DNA was effectively free to migrate between two GGG sites (a distance of ca. 7.6 nm). The stacking surface area of an ^{MD}A dimer is 1.4 times larger than that of an adenine dimer. Oxidation potential of d^{MD}A is slightly lower (−0.05 V) than that of dG. This enhanced base stacking and lower IP compared with dG plays a key role in the exceptionally high HT efficiency through the consecutive ^{MD}A sequences. Another intriguing aspect of ^{BD}A derivatives is their high stability against oxidative degradation, which will be further discussed in Section 7.4.4

7.3.2 Suppression of Hole Transport

Since HT through DNA relies on the highly ordered array of π-stacked base pairs, the suppression of HT is easily achieved by introducing perturbations into the DNA. Reduction in HT efficiency has been reported in mismatched duplexes [27], bulge structures [28], PNA-DNA hybrids [29], and DNA aptamer sequences [30]. For example, while ^{MD}A-T and G-C base pairs permit effective HT, mismatched ^{MD}A-C and G-T wobble base pairs lead to the suppression of HT [31]. These features have been applied to the DNA logic gate system.

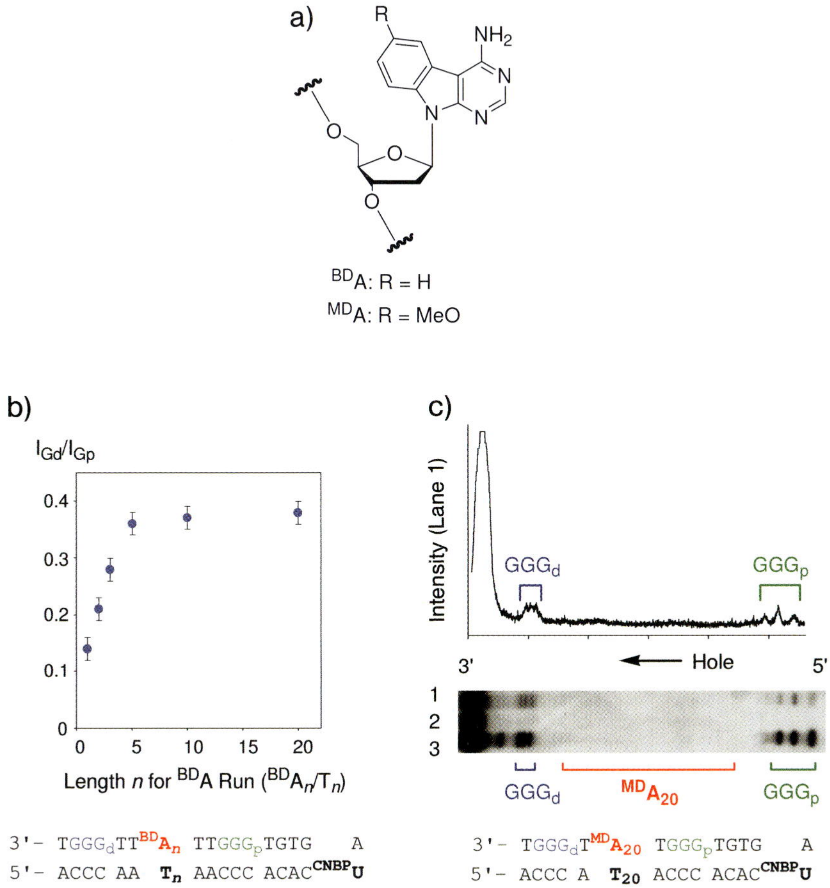

Fig. 7.4 Hole transport through ^{BD}A runs. (a) Structure of ^{BD}A derivatives. (b) Correlation of hole transport efficiency to the length of the ^{BD}A runs. (c) Densitometric analysis of the sequencing gel electrophoresis of the duplex containing $^{MD}A_{20}$. Lane 1: duplex DNA photoirradiated; lane 2: dark control; lane 3: Maxam-Gilbert G+A sequencing lane. Partial sequences used are shown at bottom.

7.3.2.1 Suppression of Hole Transport by *Bam*H I Binding

Most DNA in vivo exists as a complex with DNA-binding proteins, as typified by histones, and is quite different from the naked DNA used for our research. As the number of observations of DNA-mediated HT using ODNs increases, fundamental questions regarding the likelihood of HT in genomic DNA and its role in biology arise as the next issues to be studied [32, 33]. It has been shown that DNA π-stack disruption induced by protein binding decreases HT efficiency [34]. Besides the structural perturbations, HT is also suppressed by the binding of proteins that alter the electronic state of DNA. In particular, the distribution of the electron density on nucleotide bases is significantly altered by the hydrogen bonding of charged groups to nucleotide bases. We reveal that binding of endonuclease *Bam*H I to its recognition sequence 5′-GGATCC-3′, involving hydrogen bonding of a positively charged guanidium group to guanine, effectively suppresses the oxidation of the sequence and the HT through the binding site [35].

*Bam*H I is a restriction endonuclease that binds as a dimer to the palindromic sequence 5′-GGATCC-3′ and hydrolyses the phosphodiester linkage between the two guanines in the presence of Mg^{2+}. The X-ray structure of the *Bam*H I–DNA complex shows that direct hydrogen bonding involved in the protein-DNA contacts is condensed in a major-groove face of two G-C base pairs (Figure 7.5 a) [36]. The protein-bound DNA retains a standard B-form-like conformation without sig-

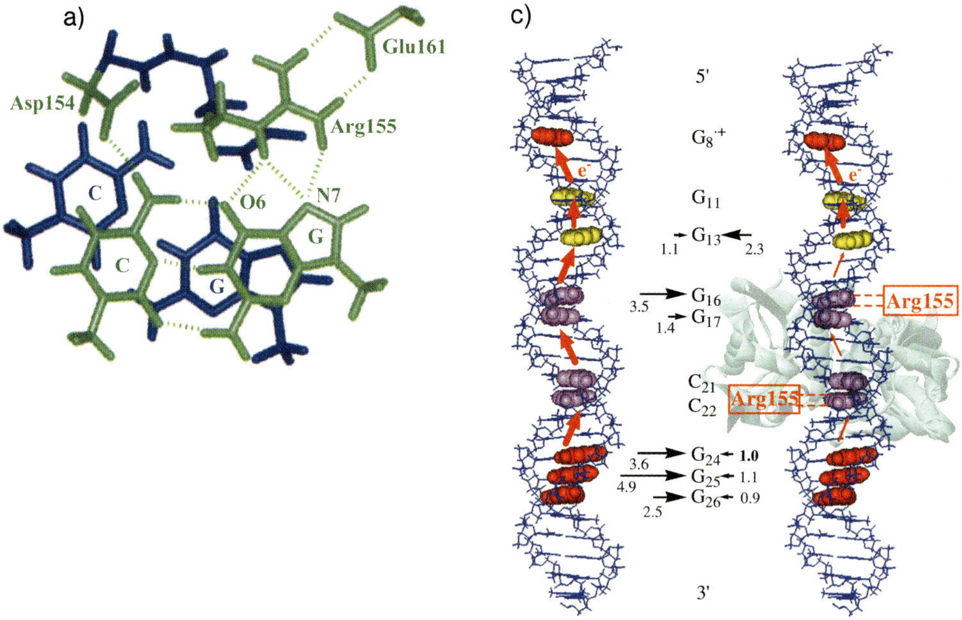

b)

```
                 8    11  13   16   21    24 2526
B1 : 5'-ATT TAT  AGT AGT GAT GGA TCC ATG GGA TTA T-3'
B2 : 3'-TAA ATA^CNBP UCA TCA CTA CCT AGG TAC CCT AAT A-5'
                                  11
```

Fig. 7.5 Hole transport through *Bam*H I–DNA complex. (a) Hydrogen bonding contacts between *Bam*H I and G-C base pairs in the recognition sequence of 5′-GGATCC-3′. The N7 and O6 atoms of 5′-side G (front side) are directly hydrogen bonded to the Arg155 guanidium group that is also bound to Glu161. The O6 of 3′-side G is bound to Asn116. For clarity, a hydrogen-bonding network involving 5′-side G-C base pairs is shown with a dotted line. (b) Sequence of ODN duplex containing a *Bam*H I recognition sequence. (c) Illustration of HT in the duplex **B1**/**B2** in the absence and presence of *Bam*H I. Horizontal arrows and the numbers shown on the site of guanine oxidation indicate the band intensity relative to that of G_{24} in the *Bam*H I–bound duplex.

nificant bends and distortions of the base stack. In contrast, the electronic states of the DNA are significantly different from the free state. Direct hydrogen bonding of Arg155 of *Bam*H I to both N7 and O6 of the 5′ guanine of the 5′-GGATCC-3′ binding motif should make the electron density of the guanine in the complex lower than in the free-state DNA. Electrostatic potentials of free guanine and guanine-Arg155-Glu161 triad in the *Bam*H I–DNA complex were calculated using DFT/B3LYP/6−31G(d) and in fact show that guanine in the complex is much more electron deficient than free guanine. Such electron-deficient guanines would become less easily oxidized because of the increase of the IP and would no longer function as a stepping-stone in HT via guanine hopping.

To address the electronic effects of protein binding on the efficiency of HT through the DNA π-stack, we examined HT in the duplex **B1**/**B2**, which contains a *Bam*H I binding site in the middle of the sequence (Figure 7.5 b). A positive charge produced at G_8 can migrate through the π-stack down to $G_{16}G_{17}$ in ODN **B1** and to $G_{11}G_{12}$ in ODN **B2**, which both directly contact a guanidium group in Arg155 of the *Bam*H I and then arrive at the hole-trapping $G_{24}G_{25}G_{26}$. Since the hole-injecting G_8 and the hole-trapping $G_{24}G_{25}G_{26}$ are located on opposite sides of the *Bam*H I binding site, the HT from G_8 to the G triplet must proceed through the site of *Bam*H I binding. Densitometric analysis of the cleavage bands of the photoirradiated duplex **B1**/**B2** are summarized in Figure 7.5c. In the absence of *Bam*H I, the hole produced at G_8 in the duplex **B1**/**B2** migrates down to $G_{24}G_{25}G_{26}$ through the $G_{16}G_{17}$ATCC sequence. Strong cleavage bands are observed at 5′G (G_{24}) and middle G (G_{25}) in the GGG hole trap, in addition to the cleavage at G_{16} of the GG site. Only a faint band was observed at G_{13} in the GA site. In the condition that ensures a complete complex formation between *Bam*H I and the duplex, strand cleavages at $G_{24}G_{25}G_{26}$ and $G_{16}G_{17}$ were significantly suppressed. Since *Bam*H I binds as a dimer to the palindromic sequence of 5′-GGATCC-3′, the $G_{11}G_{12}$ site in the complementary strand was equally insulated from one-electron oxidation. Simultaneous suppression of oxidation at both GG sites shows that insulation of both GG sites from one-electron oxidation is due to the binding of *Bam*H I to the recognition sequence. In spite of the presence of the hole-trapping G triplet, the predominant site for one-electron oxidation in the protein-bound duplex is the G_{13}A site, suggesting a considerable decrease in the efficiency of the HT from G_{13} to G_{24} (Figure 7.5 c). These results clearly show that *Bam*H I binding to DNA suppressed not only the one-electron oxidation of GG in the recognition sequence but also the HT through the protein-binding site. This remarkable effect of *Bam*H I binding on DNA-mediated HT is most ascribable to electronic insulation of the guanine in the binding site by direct contact of the positively charged guanidium group of the protein via hydrogen bonding. Contact via hydrogen bonding of a guanidium group in arginine to guanine is one of the most commonly observed protein-DNA interactions and is indispensable for the sequence recognition. Suppression of HT by the electrostatic contacts with positively charged amino residues might play a special role in the biological system.

7.4 Modulation of Hole Trapping

Once a radical cation (hole) is injected into DNA, it can migrate through the π-stacked aromatic array of DNA as described in this volume. Hole is then vanished via charge recombination or trapping reactions with oxygen and/or water. Hole trapping usually occurs at the G base, which has the lowest oxidation potential of all bases, and G damage from a distance has been used as a useful marker of HT. Remote oxidation emerges as the outcome of correlations among HT rate, back electron transfer rate, and hole trapping rate. Thus, the rate of hole trapping is clearly involved in determination of overall HT efficiency. Hole trapping is also important in the biological points of view, since it produces guanine-damaged sites that eventually cause mutations and aging. Preferential trapping at G multiplets may function as a cathodic protection against the oxidative damage of essential parts of a genome [37]. In this chapter, we will discuss our approaches for modulation of hole-trapping reactions.

7.4.1 Hole Trapping by Guanine-rich Sequences

It is widely accepted that sequences with repeated G's act as thermodynamic sinks in hole migration through the π-stacked DNA. Our group has demonstrated both experimentally and by molecular orbital calculations that stacked G's represent the most electron-donating sites in conventional B-DNA [18, 19]. The highest occupied molecular orbital (HOMO) of a GG stack is especially high in energy and is localized on the 5′G of the 5′GG sequence. These findings provide a rationale for GG-selective oxidative damage by a number of oxidants, although recent studies suggest that G multiplets are shallower thermodynamic sinks than our initial estimations would suggest [25, 38]. While the sequence effect on hole trapping has been investigated in detail, there are few studies that document the modulation of hole trapping in terms of a higher-order structure of DNA [39–42]. As examples of the structural effect on hole trapping, we will demonstrate the one-electron oxidation of the C·G*G triplex that exhibits an exceptionally efficient hole trapping at the G of the third strand of a C · G*G triplex [41].

A 26-mer pyrimidine-rich strand (**PYX**) and a 42-mer purine-rich probe strand (**PU**) is used for C · G*G triplex formation (Figure 7.6). In the presence of Mg^{2+}, the 3'-terminal purine-rich region of **PU** folds back along the major groove of the duplex with reversed Hoogsteen hydrogen bonds, producing a 12-mer sequence of an intramolecular Py-Pu-Pu triplex. Formation of a C · G*G triplex in the presence of $MgCl_2$ was confirmed by the circular dichroism spectra and dimethyl sulfate footprinting of **PYX/PU**. The results of the oxidation of the C · G*G triplex are summarized in Figure 7.6. The hybrid **PYX/PU** contains three distinct types of GGG sites, namely, $G_{12}GG$ in the duplex region, $G_{20}GG$ in the Watson-Crick purine strand of the triplex region, and $G_{35}GG$ in the third strand of the triplex. In the absence of $MgCl_2$, two GGG sites in the duplex region ($G_{12}GG$ and $G_{20}GG$) were

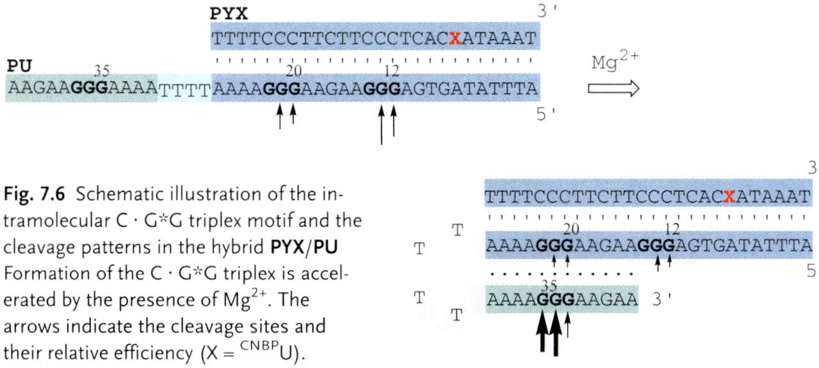

Fig. 7.6 Schematic illustration of the intramolecular C · G*G triplex motif and the cleavage patterns in the hybrid **PYX/PU** Formation of the C · G*G triplex is accelerated by the presence of Mg^{2+}. The arrows indicate the cleavage sites and their relative efficiency ($X = {}^{CNBP}U$).

cleaved via HT through DNA, whereas the cleavage at $G_{35}GG$ was not observed (Figure 7.6). The HT to $G_{35}GG$ could not occur through a long single-stranded bridge between $G_{20}GG$ and $G_{35}GG$, which would be a poor π-stacked medium. In the presence of $MgCl_2$, the cleavage efficiency at $G_{12}GG$ and $G_{20}GG$ dramatically decreased with a concomitant increase in the efficiency at $G_{35}GG$. Under these conditions, more than 90% of the cleavage is localized on $G_{35}GG$ of the third strand of the C · G*G triplex region. The hole generated at G_8 migrated to $G_{35}GG$ in the third strand via an interstrand HTand was trapped exclusively at G's in the third strand. These results clearly indicate that GGG in the third strand acts as a more effective hole trap than GGG in the double strand. This highly efficient hole trapping is interesting compared to the previous report that guanines within the $C \cdot G{*}C^+$ triplet are less reactive than G in a Watson-Crick base pair [39].

In order to elucidate the origin of the specific cleavage in the third strand of the C · G*G triplex, molecular orbital calculations on the C · G*G triplex [43] were performed at the DFT/B3LYP/6–31G level. The calculated ionization potential (IP) of $d(C \cdot G{*}G)_2$ estimated by Koopmans' theorem is 3.96 eV, which is lower than that of the duplex 5′-CCC-3′/5′-GGG-3′ (4.17 eV). It is also noteworthy that almost all of the highest occupied molecular orbital (HOMO) of $d(C \cdot G{*}G)_2$ is concentrated on the 5′-G of GG in the third strand (Figure 7.7 a). These calculations indicate that $d(C \cdot G{*}G)_2$ acts as a more effective thermodynamic sink than the GGG triplet in duplex DNA and that the 5′-G of GG in the third strand is the most electron-donating site in $d(C \cdot G{*}G)_2$.

The calculated radical cation state of the C·G*G triplex also provided significant information regarding the selective cleavage at guanines in the third strand. Comparison of optimized structure of $C_1 \cdot G_1{*}G_2$ with that of $(C_1 \cdot G_1{*}G_2)^{\bullet+}$ indicated a possible proton shift from N1 (G_2) to N7 (G_1) in the structure of $(C_1 \cdot G_1{*}G_2)^{\bullet+}$. The ease of deprotonation from G_2, not from G_1, will be critical for the selective cleavage on the third strand, because deprotonation of the guanine radical cation is considered to be the first step leading to alkali-labile products [20]. Furthermore, the spin densities of the optimized $(C_1 \cdot G_1{*}G_2)^{\bullet+}$ are completely localized on G_2 (Figure 7.7 b). Therefore, only the G_2 has an exclusively radical character in $(C_1 \cdot G_1{*}G_2)^{\bullet+}$ and can rapidly deprotonate and undergo subsequent reactions with oxygen and/or water.

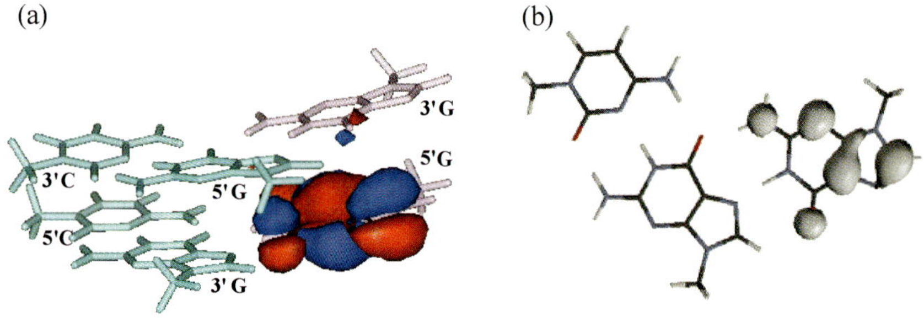

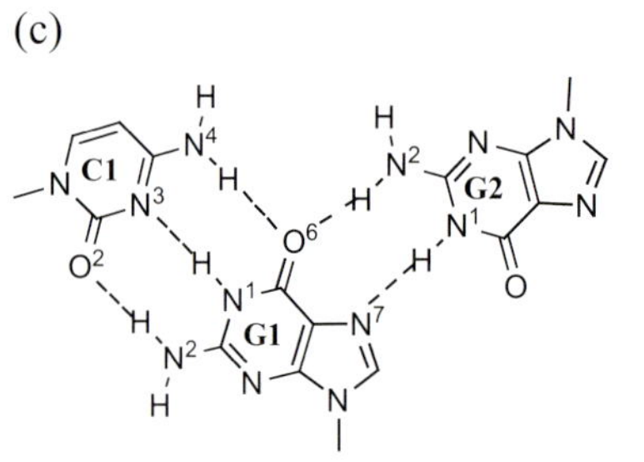

Fig. 7.7 Molecular orbital calculation of the C · G*G triplex. (a) Population density plots of the HOMO of d(C · G*G)^2 obtained by density-functional calculation with B3LYP/ 6-31G. The geometry was obtained from NMR of a 31-mer intramolecular triplex [43]. (b) Spin densities of the optimized (C · G*G)$^{\bullet+}$. For clarity, chemical structure of the C · G*G triplex is shown in (c).

Upon the formation of the C · G*G triplex, G in the third strand becomes a powerful hole trap owing to thermodynamic (lower IP) and kinetic causes (faster trapping reactions). The drastic change in hole-trapping efficiency suggests that the trapping reaction might be modulated by particular sequences and structures in biological systems [37].

7.4.2
Hole Trapping by Modified Bases

Highly efficient hole trapping is easily achieved with modified nucleosides that have lower oxidation potential than G (Figure 7.8). Naturally occurring 7,8-dihydro-8-oxo-2′-deoxyguanosine (8-oxoG), which is the preferred site of oxidation in DNA, has a lower oxidation potential [20, 44]. 8-Methoxy-2′-deoxyguanosine (8-OMeG) is the first artificial nucleoside for a hole trap in one-electron oxidation in DNA [10]. 8-OMeG acts as a thermodynamic sink of DNA-mediated HT owing to the introduction of an electron-donating methoxy group. 8-OMeG is converted effectively to an alkali-labile 2-aminoimidazolone derivative upon one-electron oxidation. From both the experimental and analytical points of view, this clean reaction is an advantage over the complex mixture of oxidation products of 8-oxoG. Other examples of hole-trapping nucleosides are the 7-deazapurine derivatives 7-deazaguanosine (^{Z}G) (Section 7.3.1.1) and 2-amino-7-deazaadenosine (^{AZ}A) [11]. ^{AZ}A is the first hole-trapping nucleoside that can form stable base pairs with both T and C. All the trapping nucleosides exhibit higher hole-trapping efficiency than the GGG triplets and have been used for the study of long-range HT.

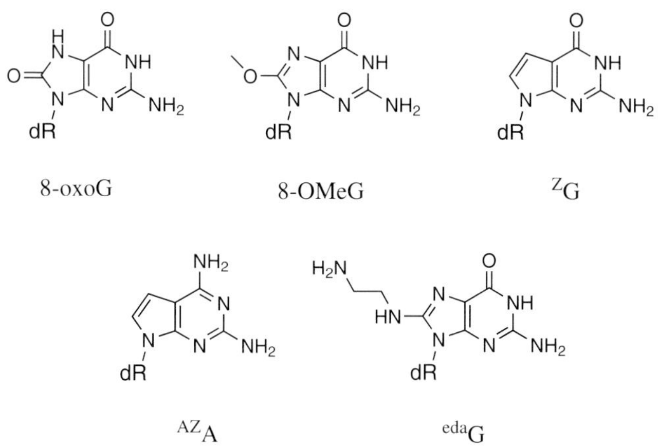

Fig. 7.8 Hole-trapping nucleosides. dR denotes 2′-deoxyribos-1′-yl.

Hole trap is usually detected as the strand cleavage after hot alkali treatment, employing the alkali-labile properties of the oxidation products of hole-trapping nucleosides. Ethylenediamine-modified G (^{eda}G) is a new type of hole-trapping nucleoside that releases a reporter molecule in response to stimulation of one-electron oxidation [12]. Hole trapping can be detected as the release of the reporter units attached at the amino side chain of ^{eda}G.

The photooxidation of benzoyl-tagged ODN, 5′-d(TAT AAT **Y** TAA TAT)-3′ (**Y** = Bz-^{eda}G), with riboflavin resulted in rapid decomposition of the ODN ($t_{1/2}$ = 6.2 min). The major isolable photoproducts of the oxidation of the benzoyl-tagged ODN were identified as ODN fragments, 3'-abasic end **2**, 5′-phosphate end **3**, and benzamide **1** possessing a guanidinium group (Scheme 7.1). The identified products strongly suggest that the rapid decomposition of ^{eda}G proceeds via the G cation radical decomposition mechanism proposed earlier [20] to result in a release of a functional unit as typically represented by **1**. Site-selective oxidation and the release of **1** were also observed with sodium hexachloroiridate (IV). Ir(IV) is a highly selective oxidant that reacts exclusively with oxidized nucleobases such as 8-oxoG and 8-oxoA. The oxidation potential ($E_{1/2}$) of Bz-^{eda}G is 0.59 V (vs. NHE),

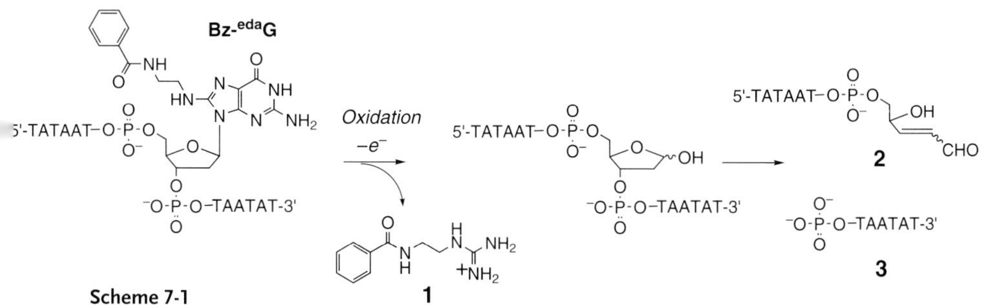

Scheme 7-1

which is close to that of 8-oxoG (0.56–0.60 V) [20]. ^{eda}G is the preferred hole-trapping site owing to its low oxidation potential.

The method using fluorophore-tethered ^{eda}G provides a simple strategy for detecting the hole trap without complicated and time-consuming analysis processes such as quantification of oxidative guanine damage of labeled DNA. We examined the detection of TAMRA released from ODN containing a TAMRA-tethered ^{eda}G via long-range HT through DNA (Figure 7.9). TAMRA-^{eda}G containing **ODN Y3** was hybridized with ^{CNBP}U-containing complementary **ODN X3** and was photoirradiated. The resulting solution was then filtered through a centrifugal filter to remove the ODNs and to collect the released fluorophore. A strong fluorescence at 576 nm was observed from the filtrate after photoirradiation, and the fluorescence after 60 min irradiation was seven times stronger than that of the control duplex without ^{CNBP}U (Figure 7.9 b). The fluorescence intensity of the **Y3/X3** sample increased in proportion to the irradiation time. The change in fluorescence intensity showed a good correlation with the strand cleavage at the ^{eda}G site, which was independently quantified by polyacrylamide gel electrophoresis (PAGE) for the experiment using **Y3/X3** (Figure 7.9 c). In addition, in the PAGE analysis for the photoirradiated duplex, it was observed that lesions at the GGG sites, located between ^{CNBP}U and ^{eda}G, were strongly suppressed. Thus, ^{eda}G acts as a very efficient hole trap, and the hole generated in the duplex by ^{CNBP}U is selectively trapped at the ^{eda}G site via a long-range HT, resulting in a release of TAMRA from the duplex.

The fluorescence from the photoirradiated sample was visually detectable. As shown in Figure 7.9 d, a strong visible emission was observable with the filtrate of the photoirradiated **Y3/X3** sample, whereas the emission from the filtrate of photoirradiated **Y3/T3**, a control sample, was negligible. The incorporation of TAMRA-^{eda}G into the duplex makes it possible to detect hole transport through DNA without PAGE analysis. This molecular releasing system, induced by stimulations of oxidation or photoirradiation, would be applicable for gene analysis and drug-releasing systems.

7.4.3
Kinetic Hole-trapping Bases

Almost all of the hole traps described in previous sections are thermodynamic hole traps. Hole injected into DNA is trapped preferably at the intended trapping sites possessing lower oxidation potentials. On the other hand, if the rate of the trapping reaction is much faster than the HT rate, the hole would be trapped primarily by a kinetic cause. Here we will discuss the kinetic hole-trapping nucleoside N^2-cyclopropyl-2′-deoxyguanosine (d^{CP}G), which possesses a cyclopropyl group at N^2 as a radical-trapping device [13].

It is known that the radical cation of cyclopropylamine and *N*-alkyl- and *N*-arylcyclopropylamines rapidly undergoes homolytic cyclopropane ring opening to produce β-iminium carbon radicals. The rate of homolytic ring opening of the cyclopropylamine radical cation is believed to be larger than that of the corresponding ring opening of the neutral *N*-alkylcyclopropylaminyl radical (7.2×10^{11} s^{-1})

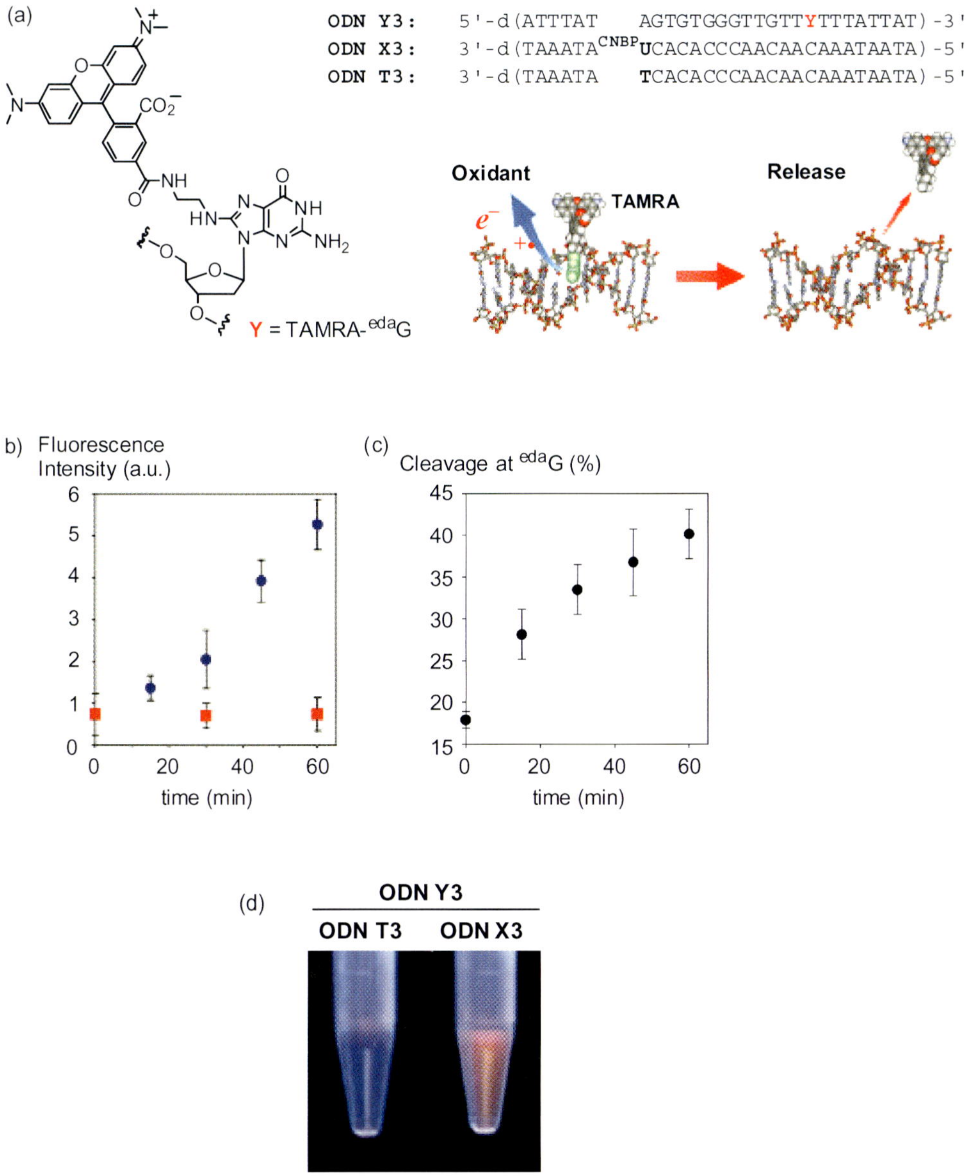

Fig. 7.9 Degradation and TAMRA release of TAMRA-^{eda}G via long-range DNA oxidation. (a) Sequences of duplex ODNs. (b) Fluorescence intensity of the reaction samples after photoirradiation and removal of ODN. Fluorescence spectra were measured at 550-nm excitation. Fluorescence intensities at 576 nm are designated by solid circles for **Y3**/**X3** duplex and by solid squares for **Y3**/**T3**. (c) Cleavage of TAMRA-^{eda}G via hole transport. The relative damaging extents show the percentage of strand breakages at the ^{eda}G site relative to the total strand cleavage obtained by densitometric analysis. (d) Fluorescence image of the samples given by the **Y3**/**T3** duplex (left) and the **Y3**/**X3** duplex (right) after 312-nm irradiation (60 min) followed by removal of ODN.

have extended this chemistry to the adenine base and developed N^6-cyclopropyl 2′-deoxyadenosine ($d^{CP}A$, Figure 7.10), which functions as a kinetic hole trap in a fashion similar to $d^{CP}G$. Detailed analysis of ^{CP}A-containing duplex shows that ^{CP}A can trap the HT process between two GG sites separated by long A/T sequences. Hole trapping at ^{CP}A strongly supports the possibility of a charge injection into long A/T-bridged sequences [14].

The rate for hole trapping, reactions of $G^{\bullet+}$ with O_2 and water, was estimated to be $6 \times 10^4\ s^{-1}$, which is significantly slower than the HT rate [46]. Oxidative damage via this slow hole-trapping pathway does not necessarily reflect the rate of hole formation and its transport in DNA [47]. While the rate constant of $d^{CP}G$ ring opening has not been determined, it is expected to be orders of magnitude faster than trapping of $G^{\bullet+}$. This fast hole trapping changes an apparent trail of DNA-mediated HT. $d^{CP}G$ can provide deeper insight into kinetic aspects of HT and direct evidence for hole generation in DNA [7, 47, 48].

7.4.4
Stable Bases Against Oxidative Degradation

Hole trapping is a serious drawback from the viewpoint of DNA as a molecular electronic device. Contrary to the hole-trapping bases, artificial bases that are not oxidatively decomposed are suitable building blocks for DNA-based electronic devices and biosensors. N^2-phenylguanosine (^{Ph}G) is the first guanine derivative that avoids oxidative degradation in keeping with HT efficiency [15].

^{Ph}G possesses a phenyl ring at the N^2-exocyclic amino group of dG, like the hole-trapping nucleoside $d^{CP}G$, which possesses a cyclopropyl group at the N^2 site. The oxidation potential of $d^{Ph}G$ is only 0.03 V higher than that of dG, and HT efficiency through the $TT^{Ph}GTT$ bridge sequence is comparable to the efficiency through the TTGTT bridge. The most remarkable difference between G and ^{Ph}G is the latter's stability against oxidative decomposition; ^{Ph}G in the duplex ODN strongly resists oxidative decomposition. The oxidative cleavage at the ^{Ph}G in 5′-^{Ph}GG-containing DNA duplex was considerably weaker (six times) than that at the G in 5′-GG-containing DNA duplex [15].

Oxidative decomposition of ^{Ph}G-containing duplex was suppressed not only at ^{Ph}G but also remarkably at the GG sites that were distant from ^{Ph}G. Band intensities at GG sites are summarized in Figure 7.12. While strand cleavage of photoirradiated **GG5** in the presence of riboflavin occurred at all GG sites with comparable efficiency, the cleavage of GG sites of **^{Ph}GG5(8)** was suppressed at $G_{12}G$ and $G_{16}G$ in addition to $^{Ph}G_8G$ (Figure 7.12 a). In contrast to $G_{12}G$, the cleavage at $G_{24}G$ was only weakly suppressed as compared to the cleavage in **GG5**, showing that the efficiency of suppression of strand cleavage decreased with increasing distance from ^{Ph}G. Distance dependency of the cleavage suppression was clearly shown in the oxidation of **^{Ph}GG5(16)** that contained ^{Ph}G in the middle of five GG sites. The efficiency of the cleavage was considerably reduced at all four GG sites in addition to the ^{Ph}GG site. Significant insights into the mechanism of cleavage suppression by ^{Ph}G were obtained by the riboflavin-sensitized oxidation of **GG4**

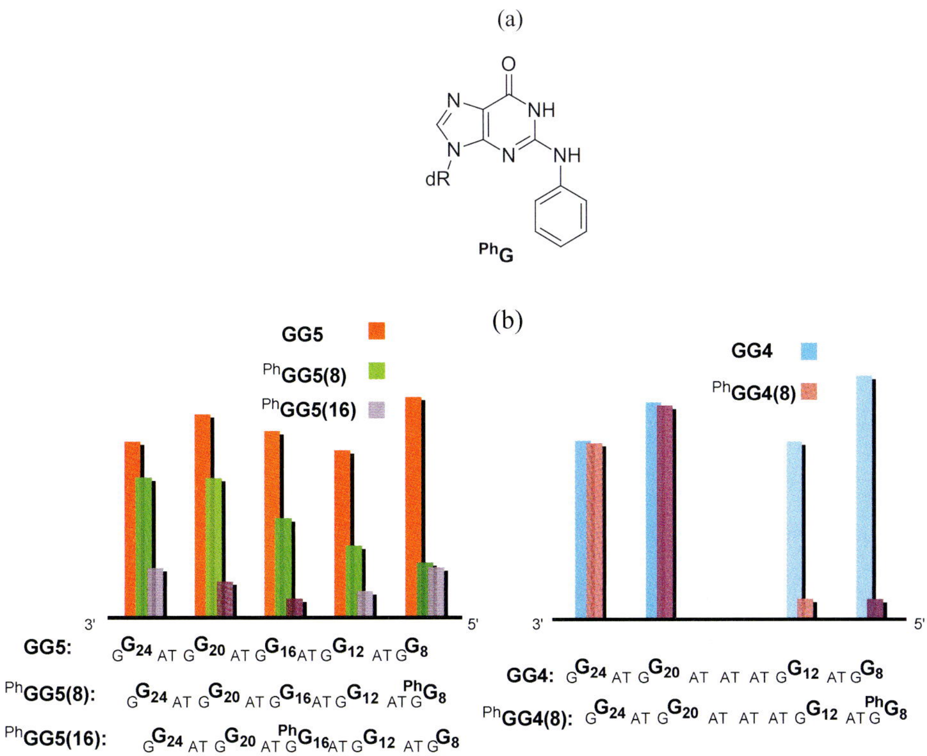

Fig. 7.12 Band intensities of GG sites relative to the intact full-length bands obtained for the photoirradiated ^{Ph}G-containing duplexes. The duplexes were irradiated in the presence of riboflavin, and the resulting oxidative damages were analyzed by a sequencing gel electrophoresis. (a) ODN **GG5**, **^{Ph}GG5(8)**, and **^{Ph}GG5(16).** (b) ODN **GG4** and **^{Ph}GG4(8).** Partial sequences are shown at bottom.

and **^{Ph}GG4(8)**. Strand cleavage of **GG4** was observed at all four GG sites, whereas the cleavage of **^{Ph}GG4**(8) was strongly suppressed at G_8G and G_{12}G sites but not at all at G_{20}G and G_{24}G sites (Figure 7.12b). The G_{12}G and G_{20}G sites in the duplex **^{Ph}GG4**(8) were separated by six intervening A-T base pairs, and the rate of HT between two sites was expected to be much smaller (Section 7.3.1.1) than that between G_8G and G_{12}G and the rate of hole trapping at the G site. Thus, it is apparent that the efficiency of suppression of the oxidative decomposition at the given G sites in ^{Ph}G-containing duplex increased with an increasing rate of HT to ^{Ph}G. Although the precise mechanism and origin of the stability of ^{Ph}G are not clear, these remarkable observations could be rationalized by assuming an annihilation process of the d^{Ph}G radical cation that was prevented from decomposing, leading to the formation of a piperidine-labile site. These results indicated that d^{Ph}G was a prototype of nucleosides functioning as an intrinsic antioxidant of duplex DNA toward one-electron oxidation.

^{BD}A derivatives (Section 7.3.1.2) have a higher resistance to oxidative degradation than does ^{Ph}G [9]. While d^{MD}A has a slightly small oxidation potential than dG, no strand cleavage band was observed for ODN containing $^{MD}A_{20}$ from the sequencing gel electrophoresis (Section 7.3.1.2, Figure 7.4c). HPLC analysis of the enzyme-digested ODN also showed almost no d^{MD}A decomposition, whereas 7% consumption of dG was observed. These results indicate that the ^{MD}A run is very stable and resistant to oxidative degradation during HT. Such an effective long-range HT without any detectable degradation of the G units is very difficult to achieve in natural DNA. The nucleophilic addition of water at the C8 position is a key reaction for the degradation of purine nucleosides [20]. ^{BD}A derivatives have a fused phenyl ring at the C7-C8 position of adenine, and this additional fused ring may suppress the oxidative damage caused by the addition of water to the resulting radical cation. Given the fact that the ^{MD}A run exhibits a remarkably high HT ability due to the enhanced base stacking (Section 7.3.1.2), ^{MD}A-containing DNA is an attractive candidate for an effective DNA wire.

Construction of nanoscale electronic devices from conducting molecules remains problematic due to the difficulties of achieving inter-element wiring and electrical interfacing of molecular wires to macroscopic electrodes. One of the special features of DNA is a highly selective recognition ability with its complementary DNA, which is a distinct advantage over known conductive materials such as carbon nanotubes and conductive polymers. The remarkable recognition properties of DNA enable site-specific fabrication and functionalization of DNA on a molecular scale. DNA-based molecular wires are expected to constitute a new nanomaterial that will be widely applicable to electronic devices and biosensors.

7.5
Conclusions

We have used a rational design to produce artificial bases that have diverse and distinct chemical properties compared to natural bases. The designed bases function as a useful tool for studies of DNA-mediated HT. In addition, incorporation of the specially designed bases enables us to modulate the efficiency of hole injection, hole transport, and hole trapping. Efforts to modulate HT efficiency will provide useful information for understanding the mechanisms and biological consequences of HT as well as for applying the DNA-based electronic devices and biosensors.

References

1 S. Delaney, J. K. Barton, *J. Org. Chem.* **2003**, *68*, 6475–6483.
2 G. B. Schuster, *Acc. Chem. Res.* **2000**, *33*, 253–260.
3 F. D. Lewis, R. L. Letsinger, M. R. Wasielewski, *Acc. Chem. Res.* **2001**, *34*, 159–170.
4 B. Giese, *Annu. Rev. Biochem.* **2002**, *71*, 51–70.
5 K. Nakatani, C. Dohno, I. Saito, *J. Am. Chem. Soc.* **1999**, *121*, 10854–10855.
6 K. Kawai, T. Takada, S. Tojo, N. Ichinose, T. Majima, *J. Am. Chem. Soc.* **2001**, *123*, 12688–12689.
7 T. T. Williams, C. Dohno, E. D. A. Stemp, J. K. Barton, *J. Am. Chem. Soc.* **2004**, *126*, 8148–8158.
8 K. Nakatani, C. Dohno, I. Saito, *J. Am. Chem. Soc.* **2000**, *122*, 5893–5894.
9 A. Okamoto, K. Tanaka, I. Saito, *J. Am. Chem. Soc.* **2003**, *125*, 5066–5071.
10 H. Ikeda, I. Saito, *J. Am. Chem. Soc.* **1999**, *121*, 10836–10837.
11 A. Okamoto, K. Tanaka, I. Saito, *Bioorg. Med. Chem. Lett.* **2002**, *12*, 3641–3643.
12 A. Okamoto, K. Tanaka, I. Saito, *J. Am. Chem. Soc.* **2004**, *126*, 416–417.
13 K. Nakatani, C. Dohno, I. Saito, *J. Am. Chem. Soc.* **2001**, *123*, 9681–9682.
14 C. Dohno, A. Ogawa, K. Nakatani, I. Saito, *J. Am. Chem. Soc.* **2003**, *125* 10154–10155.
15 K. Nakatani, C. Dohno, I. Saito, *J. Am. Chem. Soc.* **2002**, *124*, 6802–6803.
16 K. Nakatani, C. Dohno, I. Saito, *J. Org. Chem.* **1999**, *64*, 6901–6904.
17 K. Nakatani, C. Dohno, T. Nakamura, I. Saito, *Tetrahedron Lett.* **1998**, *39*, 2779–2782.
18 I. Saito, M. Takayama, H. Sugiyama, K. Nakatani, A. Tsuchida, M. Yamamoto, *J. Am. Chem. Soc.* **1995**, *117*, 6406–6407.
19 H. Sugiyama, I. Saito, *J. Am. Chem. Soc.* **1996**, *118*, 7063–7068.
20 C. J. Burrows, J. G. Muller, *Chem. Rev.* **1998**, *98*, 1109–1151.
21 K. Nakatani, C. Dohno, I. Saito, *Tetrahedron Lett.* **2000**, *41*, 10041–10045.
22 F. D. Lewis, R. S. Kalgutkar, Y. Wu, X. Liu, J. Liu, R. T. Hayes, S. E. Miller, M. R. Wasielewski, *J. Am. Chem. Soc.* **2000**, *122*, 12346–12351.
23 I. Saito, T. Nakamura, K. Nakatani, Y. Yoshioka, K. Yamaguchi, H. Sugiyama, *J. Am. Chem. Soc.* **1998**, *120*, 12686–12687.
24 A. A. Voityuk, N. Rösch, *J. Phys. Chem. B* **2002**, *106*, 3013–3018.
25 M. Bixon, J. Jortner, *J. Am. Chem. Soc.* **2001**, *123*, 12556–12567.
26 S. V. Rakhmanova, E. M. Conwell, *J. Phys. Chem. B* **2001**, *105*, 2056–2061.
27 P. K. Bhattacharya, J. K. Barton, *J. Am. Chem. Soc.*, **2001**, *123*, 8649–8656.
28 D. B. Hall, J. K. Barton, *J. Am. Chem. Soc.* **1997**, *119*, 5045–5046.
29 A. Okamoto, K. Tanabe, C. Dohno, I. Saito, *Bioorg. Med. Chem.* **2002**, *10*, 713–718.
30 R. P. Fahlman, D. Sen, *J. Am. Chem. Soc.* **2002**, *124*, 4610–4616.
31 A. Okamoto, K. Tanaka, I. Saito, *J. Am. Chem. Soc.* **2004**, *126*, 9458–9463.
32 E. M. Boon, A. L. Livingston, N. H. Chmiel, S. S. David, J. K. Barton, *Proc. Natl. Acad. Sci. U.S.A.* **2003**, *100*, 12543–12547.

33 M. E. Núñez, K. T. Noyes, J. K. Barton, *Chem. Biol.*, **2002**, *9*, 403–415.
34 S. R. Rajski, J. K. Barton, *Biochemistry*, **2001**, *40*, 5556–5564.
35 K. Nakatani, C. Dohno, I. Saito, *Chem. Biol.* **2002**, *9*, 361–366.
36 M. Newman, T. Strzelecka, L. F. Dorner, I. Schildkraut, A. K. Aggarwal, *Science* **1995**, *269*, 656–663.
37 K. A. Friedman, A. Heller, *J. Phys. Chem. B* **2001**, *105*, 11859–11865.
38 F. D. Lewis, T. Wu, X. Liu, J. Liu, R. T. Hayes, M. R. Wasielewski, *J. Am. Chem. Soc.* **2000**, *122*, 12037–12038.
39 Y. Kan, G. B. Schuster, *J. Am. Chem. Soc.* **1999**, *121*, 11607–11614.
40 D. T. Odom, E. A. Dill, J. K. Barton, *Nucleic. Acids Res.*, **2001**, *29*, 2026–2033.
41 C. Dohno, K. Nakatani, I. Saito, *J. Am. Chem. Soc.* **2002**, *124*, 14580–14585.
42 S. Delaney, J. K. Barton, *Biochemistry*, **2003**, *42*, 14159–14165.
43 I. Radhakrishnan, J. D. Patel, *Structure*, **1993**, *1*, 135–152.
44 S. M. Gasper, G. B. Schuster, *J. Am. Chem. Soc.* **1997**, *119*, 12762–12771.
45 O. M. Musa, J. H. Horner, H. Shahin, M. Newcomb, *J. Am. Chem. Soc.* **1996**, *118*, 3862–3868.
46 B. Giese, M. Spichty, *Chem. Phys. Chem.* **2000**, *1*, 195–198.
47 C. Dohno, E. D. A. Stemp, J. K. Barton, *J. Am. Chem. Soc.*, **2003**, *125*, 9586–9587.
48 M. A. O'Neill, C. Dohno, J. K. Barton, *J. Am. Chem. Soc.* **2004**, *126*, 1316–1317.

8
Spectroscopic Investigation of Charge Transfer in DNA

Vladimir Shafirovich and Nicholas E. Geacintov

8.1
Introduction

In living systems, DNA molecules undergo continuous attack by free radicals and by other oxidizing species. These reactions give rise to diverse, potentially mutagenic, oxidative modifications (lesions) of the nucleobases. A significant body of experimental evidence has been accumulated showing that the formation of DNA lesions occurs not only at the DNA bases targeted by the oxidizing agents but also at nucleobases remote from these sites [1–3]. These effects have been called DNA chemistry at a distance [4]. The oxidative transformation of normal nucleobases to oxidized products involves a series of complex steps beginning from the initial electron abstraction step, the migration of a mobile intermediate (radical cation or neutral radical) along the DNA helix, and the trapping of this intermediate at a particular DNA base, followed by its chemical transformation to a stable end product. The primary target of oxidative modification in DNA is guanine, which is the most easily oxidizable nucleobase among the natural DNA bases (A, C, and T) [5].

Our group has focused on the direct spectroscopic investigations of guanine oxidation at a distance in double-stranded oligo-2'-deoxyribonucleotides and on the formation of the end products of guanine oxidation. A laser flash photolysis method developed by our group was employed for the site-specific injection of oxidizing species into the DNA double helix [6–11]. This method involves the site-specific incorporation of a single 2-aminopurine (2AP) residue, a nucleic acid base analogue, into oligonucleotide strands containing guanine residues. The selective photoionization of the 2AP residues with intense 308-nm excimer laser pulses by a two-photon absorption mechanism yields 2AP radicals that subsequently and selectively oxidize nearby guanine residues. The remote oxidation of guanine residues by 2AP radicals was monitored by transient absorption spectroscopy. The end products of guanine oxidation were isolated by a combination of reversed-phase HPLC and identified by mass spectrometry methods.

We will begin with a description of the methods we developed for the site-selective injection of oxidizing species into double-stranded DNA [6–11].

Charge Transfer in DNA: From Mechanism to Application. Edited by H.-A. Wagenknecht

ISBN: 3-527-31085-1

we estimate that, in water, the deprotonation rate constant of $2AP^{\bullet+}$ is ~3×10^7 s^{-1}, i.e., the lifetime of the free base $2AP^{\bullet+}$ is about 30 ns [6]. The characteristic time of the $2AP^{\bullet+}$ deprotonation in DNA duplexes is expected to be similar to that of free base $2AP^{\bullet+}$ (~30 ns) [6]. Indeed, the H2 proton in $2AP^{\bullet+}$ does not participate in Watson-Crick hydrogen bonding with the complementary T base; therefore, the $2AP^{\bullet+}$ deprotonation does not require a base pair opening event as in the case of the exchange of imino protons with solvent or base of Watson-Crick base pairs in double-stranded DNA [30–32]. Since the $2AP^{\bullet+}$ radical cations in DNA duplexes are also reduced by oxidative electron transfer reactions with guanine, their lifetimes are likely to be shorter than 30 ns. Therefore, we assume that in DNA duplexes, the decay of $2AP^{\bullet+}$ radical cations generated by the 308-nm laser pulses occurs mostly via two competitive pathways: (1) deprotonation of $2AP^{\bullet+}$ with the formation of the neutral radical, $2AP(\text{-}H)^{\bullet}$, and (2) selective oxidation of guanine residues by $2AP^{\bullet+}$.

The neutral $2AP(\text{-}H)^{\bullet}$ radical is a strong one-electron oxidant that selectively oxidizes guanine bases in DNA [6–8, 10]. The time window for the observation of guanine oxidation by $2AP(\text{-}H)^{\bullet}$ at a distance is expected to be in the range of ~30 ns (deprotonation of $2AP^{\bullet+}$) to ~0.5 ms (decay of $2AP(\text{-}H)^{\bullet}$ in side reactions). The existence of two potential one-electron oxidants ($2AP^{\bullet+}$ and $2AP(\text{-}H)^{\bullet}$) with very different lifetimes and reactivities toward guanine suggests that heterogeneous decay kinetics of guanine radical formation in DNA might be observable on nanosecond to microsecond time scales. However, our experiments reveal several different time-dependent and sequence-dependent kinetic phases of guanine radical formation in the observation window beginning from 100 ns (time resolution of our laser flash photolysis setup in a 300–400-nm spectral range) to 0.5 ms or more (decay of $G(\text{-}H)^{\bullet}$ radicals formed). These observations suggest that other factors may give rise to non-exponential electron transfer kinetics [7, 8, 10].

8.3.3. Spectroscopic Monitoring of the One-electron Oxidation of Guanine at a Distance

An example of the transient absorption spectra of the 2AP-modified duplexes is shown in Figure 8.1. At a delay time of 100 ns between the excitation and first transient absorbance observation time point, the signal is attributed to the superposition of the spectra of the $2AP(\text{-}H)^{\bullet}$ and the $G^{\bullet+}/G(\text{-}H)^{\bullet}$ radicals and the hydrated electrons (Figure 8.1A). The structureless tail of the e_h^- absorption in the 500–700-nm region (observed at Δt=100 ns) decays completely within the time interval of <300 ns as shown by the 300-ns spectrum. The formation of $G^{\bullet+}/G(\text{-}H)^{\bullet}$ radicals monitored by the rise of the 315-nm absorption band and associated with the decay of the $2AP^{\bullet+}/2AP(\text{-}H)^{\bullet}$ transient absorption bands at 365–510 nm occurs in at least three well-separated time domains (Figure 8.1 B, C). The prompt ($\leq$100 ns) rise of the transient absorbance at 315 nm due to the one-electron oxidation of guanine by $2AP^{\bullet+}$ was not resolved in our experiments (Figure 8.1B). The $2AP^{\bullet+}/2AP(\text{-}H)^{\bullet}$ radicals do not absorb near 315 nm [29], and the initial sig-

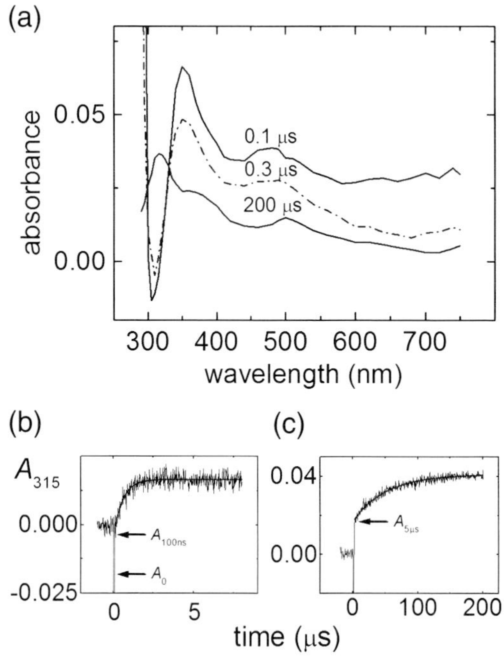

Fig. 8.1 (a) Transient absorption spectra of the [2AP]A_4GGA_8 duplex (0.1 mM) in oxygenated 20 mM phosphate buffer (pH 7) solutions recorded after 308-nm XeCl excimer laser pulse excitation (12 ns, 60 mJ pulse^{-1} cm^{-2}). (b, c) Kinetics of $G^{\bullet+}$/G(-H)$^{\bullet}$ formation in the [2AP]A_4GGA_8 duplex recorded at 315 nm in the two different time domains indicated. Solid lines are best single-exponential fits to the experimental data points. (Reproduced with permission from *Top. Curr. Chem* **2004**, *237*, 129–157. Copyright 2004, Springer-Verlag, Berlin, Heidelberg).

nal associated with the photoionization of 2AP within a laser pulse (~12 ns) is expected to be negative due to depopulation of the 2AP ground state (bleaching of the 2AP absorption band). The amplitude (A_0) of this signal (Figure 8.1B) can be assessed from the 315-nm kinetic profile of the controlled [2AP]A_{14} duplex containing no guanine [7]. The expected rise of the signal amplitude from A_0 to A_{100ns} (starting point of the observation) was not time resolved because of the scattered laser light and occurs within the response time of our setup (Figure 8.1B). In addition to this fast component (< 100 ns), two slower components (~0.5 μs in Figure 8.1B, and 60 μs in Figure 8.1C) are evident in the time-resolved kinetics of the formation of the oxidized species $G^{\bullet+}$/G(-H)$^{\bullet}$ in the microsecond and millisecond time domains. Thus, the kinetics of formation of the one-electron oxidation products, $G^{\bullet+}$/G(-H)$^{\bullet}$, in the 2AP-modified duplexes is highly heterogeneous in character.

Analysis of the transient absorption spectra allows for the determination of the relative yields of the $G^{\bullet+}$/G(-H)$^{\bullet}$ radicals at delay times of $\geq$100 ns attributed to the oxidation of the guanines by $2AP^{\bullet+}/2AP(\text{-}H)^{\bullet}$ radicals [7, 10]. The prompt yield at the 100-ns observation time point, Φ_{100ns}, associated for the most part with the oxidation of the guanines by $2AP^{\bullet+}$ radical cations, may be expressed as follows:

$$\Phi_{100ns} = [G^{\bullet+}/G(\text{-}H)^{\bullet}]_{100ns}/([G^{\bullet+}/G(\text{-}H)^{\bullet}]_{100ns} + [2AP(\text{-}H)^{\bullet}]_{100ns}) \quad (3)$$

where $[G^{\bullet+}/G(\text{-}H)^{\bullet}]_{100ns}$ and $[2AP(\text{-}H)^{\bullet}]_{100ns}$ are the concentrations of the radicals measured at a delay time of 100 ns. At this time point, all of the $2AP^{\bullet+}$ radical cations most likely have been converted to the less reactive neutral radical $2AP(\text{-}H)^{\bullet}$ via deprotonation of $2AP^{\bullet+}$ radical cations, or have been reduced back to the starting 2AP by oxidative electron transfer from guanine. The rather sharp transition in the transient absorption kinetics observed at 315 nm (resulting from the formation of $G^{\bullet+}/G(\text{-}H)^{\bullet}$ radicals) near 100 ns suggests that the $2AP^{\bullet+}$ radical cation is a more efficient oxidant than the neutral $2AP(\text{-}H)^{\bullet}$ radical. The values of the initial yield of $G^{\bullet+}/G(\text{-}H)^{\bullet}$ radicals measured at a delay time of 100 ns decrease as the number of bridging T or A bases is increased (Figure 8.2). In the $[2AP]A_nGGA_{12-n}$ duplexes, the decrease in Φ_{100ns} as a function of the increasing number of adenine bridging bases is much less pronounced than in the case of the duplexes with the $[2AP]T_nGGT_{12-n}$ strand with intervening T-bridging residues, and Φ_{100ns} approaches zero in the duplex with six bridging adenine bases (n = 6). At a given fixed number of bases (n), a greater value of Φ_{100ns} is indicative of a greater rate of oxidative electron transfer from guanine to $2AP^{\bullet+}$ radical cations. Oxidation of guanine by $2AP^{\bullet+}$ can be considered as a hole transfer from $2AP^{\bullet+}$ to guanine. Enhancement of the hole transfer reaction rate through the adenine-bridging sequence in comparison with the thymine-bridging sequence is consistent with the data of Lewis, Wasielewski, Majima, and their coworkers obtained by laser kinetic spectroscopy studies of electron transfer from guanine to diverse electron acceptors [33–36].

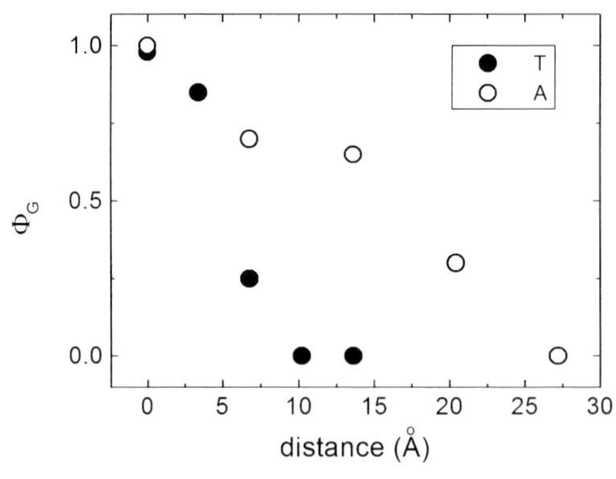

Fig. 8.2 The prompt ($\Delta t = 100$ ns) yields of the $G^{\bullet+}/G(\text{-}H)^{\bullet}$ radicals, Φ_{100ns}, as a function of the number of bridging bases in double-stranded oligonucleotides (using data from [7–10]).

8.3.4
Base Sequence Effects on the Rates of One-electron Oxidation of Guanine

The deprotonation of $2AP^{\bullet+}$ radical cations does not seem to significantly hinder the one-electron oxidation of guanine at a distance. The neutral radical, $2AP(\text{-H})^{\bullet}$, resulting from a proton release from $2AP^{\bullet+}$, remains a strong one-electron oxidant that can oxidize guanine at the level of free nucleosides [6, 29, 37] and in single- and double-stranded DNA [6–8, 10]. In the case of the $[2AP]A_6GGA_6$ and $[2AP]A_8GGA_4$ duplexes with six and eight bridging adenine bases, the transient absorption measurements indicate that guanine oxidation by $2AP(\text{-H})^{\bullet}$ is dominant and that the electron transfer rate constants are of the same order of magnitude as in the $[2AP]T_2GGT_{10}$ and $[2AP]T_3GGT_9$ duplexes, which have only two and three bridging thymidines, respectively [7, 8, 10].

The kinetics of the transient absorption signals, $A(t)$, of $G(\text{-H})^{\bullet}$ radicals recorded at 315 nm can be represented by the following equation:

$$A(t) = A \exp(-k_1 t) + B \exp(-k_2 t) \tag{4}$$

where k_1 is associated with the growth of the signal (at $t > 5\ \mu s$) due to the oxidation of guanines by $2AP(\text{-H})^{\bullet}$ radicals (Figure 8.1C) and k_2 is the observed rate constant of decay of the $G(\text{-H})^{\bullet}$ radicals in subsequent reactions ($t > 0.5$ ms). The values of k_1 were found to be independent of the concentration of the duplexes in the 10–100 μM range, as expected for intraduplex electron transfer processes. The rate constant, k_1 is composed of two terms, k_a and k_{ag}, with $k_1 = k_a + k_{ag}$. We denote the rate constant of guanine oxidation by k_{ag}, while the decay of the $2AP(\text{-H})^{\bullet}$ radical by other pathways (not involving electron transfer from G) is denoted by k_a. The rate constant k_a was measured using $[2AP]A_{14}$ and $[2AP]T_{14}$ duplexes; in these duplexes, the $2AP(\text{-H})^{\bullet}$ radicals cannot decay by direct electron transfer reactions since there are no guanines ($k_{ag} = 0$). The kinetic parameters determined by the best fits of Eq. (4) to the experimental transient absorption profiles are summarized in Table 8.1.

The effects of the number of intervening bases on the rate constant of guanine oxidation by the $2AP(\text{-H})^{\bullet}$ radicals is described by the following equation:

$$k_{ag} = k_{ag}^{o} \exp(-\beta r) \tag{5}$$

where β is an attenuation parameter and r is the distance between electron donor and acceptor residues. Plots of the linearized form of Eq. (5) using the available experimental data points for the duplexes with $n \geq 2$ intervening bases are shown in Figure 8.3. Data [9] for the oxidation of 8-oxo-7,8-dihydroguanine (8-oxoGua) by the $2AP(\text{-H})^{\bullet}$ radical in the $[2AP]T_n[8\text{-oxoGua}]T_{13-n}$ duplexes ($n = 2$ and $n = 4$) are also shown in Figure 8.3. In these duplexes with T-bridging bases, the oxidation of G or 8-oxoGua by the $2AP(\text{-H})^{\bullet}$ radicals is the slow but major pathway for the formation of the $G(\text{-H})^{\bullet}$ radicals. The slopes of these plots are a measure of the parameter β. These values depend on the sequence and do not depend on the values of k_{ag}. For instance, the absolute values of k_{ag} for oxidation of 8-oxo-dG are

Table 8.1 Deuterium isotope effect on the rate constants (k_{ag}) of the oxidation of guanine and 8-oxoguanine by 2-aminopurine neutral radicals in double-stranded DNA duplexes [7–10].

Sequence[a)]	k_{ag} (10^3 s^{-1})		
	H_2O	D_2O	H_2O/D_2O
[2AP]$TGGT_{11}$	500 $\pm$50[b)]		
[2AP]T_2GGT_{10}	10.3$\pm$1	5.9$\pm$0.6	1.7$\pm$0.2
[2AP]T_3GGT_9	3.3$\pm$0.3	2.2$\pm$0.2	1.5$\pm$0.2
[2AP]T_4GGT_8	0.9$\pm$0.1	n.d.[c)]	n.d.
[2AP]T_2GT_{11}	9.9$\pm$1	5.7$\pm$0.6	1.7$\pm$0.2
[2AP]T_2GGGT_9	17.9$\pm$2	11.8$\pm$1.2	1.5$\pm$0.2
[2AP]A_2GGA_{10}	24.2$\pm$2	12.7$\pm$1.3	1.9$\pm$0.2
[2AP]A_4GGA_8	13.0$\pm$2	7.7$\pm$0.8	1.9$\pm$0.2
[2AP]A_6GGT_6	6.0$\pm$0.6	4.6$\pm$0.5	1.3$\pm$0.1
[2AP]A_8GGA_4	2.2$\pm$0.2	1.7$\pm$0.2	1.3$\pm$0.1
[2AP]T_2[8-oxo-dG]T_{10}	38 $\pm$5	n.d.	n.d.
[2AP]T_4[8-oxo-dG]T_8	3.0$\pm$0.5	n.d.	n.d.

a) Oligodeoxyribonucleotide sequences are written in the 5′ to 3′ direction. The fully complementary strands with T opposite the 2AP residue are not shown.
b) The uncertainties are given as standard errors of the best least-squares fits of the appropriate kinetic equations to the transient absorption profiles of the G(-H)$^•$ decay curves recorded at 315 nm.
c) n.d. = not determined.

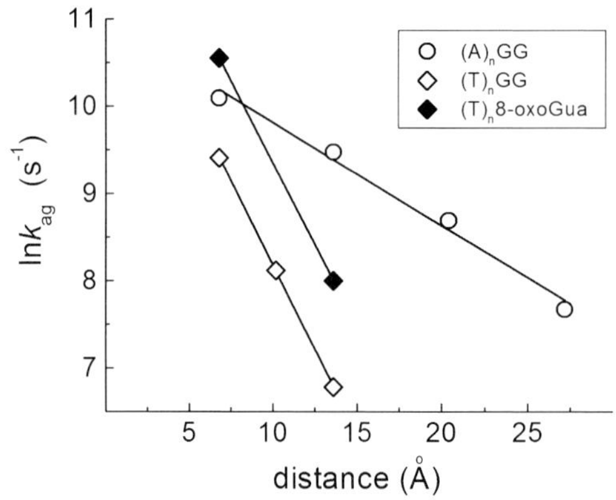

Fig. 8.3 Rate constants of proton-coupled electron transfer from guanine to 2AP(-H)$^•$ radicals, k_{ag}, as a function of the number of bridging bases in double-stranded oligonucleotides. (Reproduced with permission from *Top. Curr. Chem* **2004**, *237*, 129–157. Copyright 2004, Springer-Verlag, Berlin, Heidelberg).

greater than those for the oxidation of guanine. Nevertheless, based on the only two available data points for the sequences with 8-oxo-dG, the distance dependency of the duplexes with T sequences appears to be characterized by the same parameter $\beta = 0.4$ Å^{-1} as the sequences with guanine as the electron donor. Note that a greater value of $\beta = 0.75$ Å^{-1} was previously obtained using the data for the duplexes with one, two, and three T-bridging residues [7]. In the duplexes with A sequences, the distance dependence is lower since $\beta = 0.12$ Å^{-1}. Our β parameters are in a good agreement with the data of Lewis, Wasielewski, and Majima and their coworkers, also determined by time-resolved spectroscopic methods [33–36, 38]. There is general agreement, based on direct measurements of electron transfer rates in different oligonucleotide duplexes, that A-bridging sequences are more efficient than the T-bridging sequences in mediating the one-electron oxidation of guanine not only by 2AP radical cations but also by neutral 2AP radicals.

The oxidation of guanine by 2AP(-H)$^{\bullet}$ radicals depends only weakly on the number of contiguous guanines in the sequence context 5′-...G..., 5′-...GG..., and 5′-...GGG.... The ratios of the electron transfer rate constant, k_{ag}, in the sequences [2AP]T2$G_m T_{12-m}$ where $m = 1$ (single G), 2 (..GG.), or 3 (...GGG...) are 1:1.04:1.8 (Table 8.1). In the case of oxidation of guanines by photoexcited stilbene, the analogous ratios were 1:1.7:1.5, respectively [39], and thus are comparable to our results. The estimates by Lewis et al. [89] have shown that even these small differences in the rate constants can lead to modest differences in the photochemical oxidation of guanines as measured by the alkali labile strand cleavage–gel electrophoresis methods [40–47].

8.3.5
Proton-coupled Electron Transfer at a Distance

A kinetic deuterium isotope effect on the rates of oxidation of guanine by 2AP radicals in double-stranded DNA [8, 10] showed that this reaction can be considered in terms of a proton-coupled electron transfer [37]. The values of k_{ag} in H_2O buffer solutions are larger than those in D_2O buffer solutions by factors of 1.3–1.7 (Figure 8.4). The magnitude of the kinetic isotope effect, defined by the ratio $k_{ag}(H_2O)/k_{ag}(D_2O)$, decreases somewhat as a function of an increasing number of bridging bases (Table 8.1), as predicted by theoretical considerations of proton-coupled electron transfer reactions occurring at fixed distances [48]. In the DNA duplexes, the electron transfer from G to 2AP(-H)$^{\bullet}$ is coupled to a deprotonation of the radical cation $G^{\bullet+}$ and a protonation of the anion 2AP(-H)$^{-}$. In B-form DNA, the H2 sites of 2AP are accessible and the protonation of the 2AP(-H)$^{-}$ anion does not require any change in the hydrogen-bonding configuration between the nucleic acid bases in the duplex.

The free base $dG^{\bullet+}$ radical cation is a Brønsted acid ($pK_a = 3.9$) [27] and should rapidly deprotonate with the rate constant k_{-d} in neutral solutions of free nucleosides:

$$dG^{\bullet+} \underset{k_d}{\overset{k_{-d}}{\rightleftharpoons}} dG(\text{-H})^{\bullet} + H^+ \qquad (6)$$

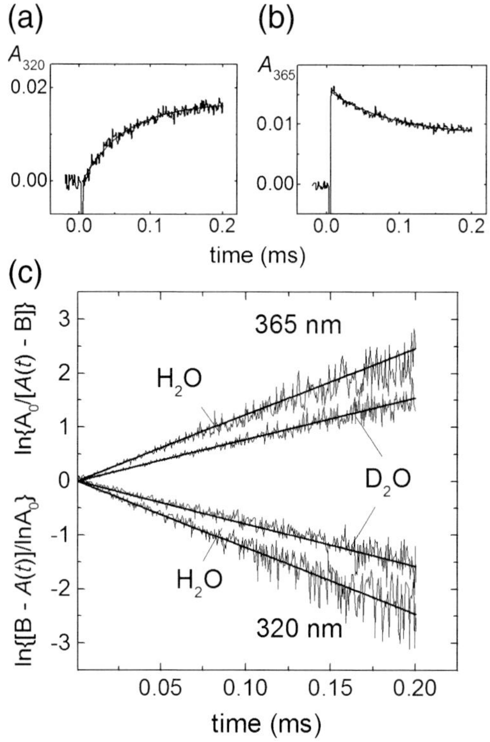

Fig. 8.4 Deuterium isotope effect on the kinetics of oxidation of G by 2AP(-H)$^{\bullet}$ radicals in the [2AP]T_2GGT_{10} duplex in oxygenated H_2O/D_2O buffer solutions (pH 7.0). The kinetic profiles (resolution of 0.5 s per point) of 2AP(-H)$^{\bullet}$ decay (365 nm) and G(-H)$^{\bullet}$ formation (320 nm) (a and b) were linearized according to a semi-logarithmic form of Eq. (4) as shown in (c). Solid lines are the best linear fits to the experimental data. (Reproduced with permission from *J. Phys. Chem. B* **2001**, *105*, 8431–8435. Copyright 2001, American Chemical Society).

Employing a pulse radiolysis technique, Kobayashi and Tagawa obtained the value of $k_{-d} = \sim 1.8 \times 10^7\ s^{-1}$ for deprotonation of the free base $dG^{\bullet +}$ radical cation in 10 mM phosphate buffer solution (pH 7) [49]. In double-stranded oligonucleotides, the kinetics of $G^{\bullet +}$ deprotonation was heterogeneous and includes a fast ($\sim 1.3 \times 10^7\ s^{-1}$) and a slow ($\sim 3.0 \times 10^6\ s^{-1}$) component that were similar for different oligonucleotides with single G, GG, or GGG steps. However, in oligonucleotides containing single G residues located either at or near the 5′- and 3′-terminal positions, only the fast ($\sim 1.3 \times 10^7\ s^{-1}$) component was detected, thus demonstrating that the lifetimes of $G^{\bullet +}$ radical cations depend on their positions in DNA duplexes. The heterogeneous kinetics of the $G^{\bullet +}$ deprotonation in DNA are probably associated with a multi-step mechanism of the deprotonation process involving a shift of the N1 proton in $G^{\bullet +}$ to its Watson-Crick partner cytosine, followed by a final release of the proton into solution. Nevertheless, even in double-stranded

DNA, deprotonation of $G^{\bullet+}$ radical cations remains very fast, and the lifetime of the slowest component (~300 ns) is only a factor of six longer than that of the free base $dG^{\bullet+}$ radical cation (~60 ns) [49].

The solvent kinetic deuterium isotope effect on the rates of guanine oxidation is typical of its reactions with other one-electron oxidants. Thus, the reactions of the radical cation of a pyrene derivative, 7,8,9,10-tetrahydroxytetrahydrobenzo[a]pyrene (BPT), or of the noncovalent complex of $BPT^{\bullet+}$ with dAMP exhibit similar deuterium isotope effects [50]. The kinetic isotope effects observed for photoinduced electron transfer reactions in covalently linked benzo[a]pyrene diol epoxide–guanosine adducts [51], noncovalent benzo[a]pyrenetetrol-nucleoside complexes [52], DNA-intercalating ruthenium complexes [53], electrocatalytic oxidation of guanine in DNA [54], and excess electron migration in DNA [55] provide strong support for the notion that these reactions are all coupled with a proton transfer process. Experimental and theoretical studies showed that proton-coupled electron transfer reactions can occur in DNA-acrylamide complexes [56, 57], in the photooxidation occurring in noncovalent guanine-cytosine complexes by *N,N'*-dibutyl-1,4,5,8-naphthalenediimide or fullerene in the triplet excited states [58, 59], and in the oxidation of phenol by guanine radicals in plasmid DNA [60].

8.4
End Products of Guanine Oxidation

In neutral aqueous solutions, the ultimate product of one electron abstraction from guanine is the guanine neutral radical. In DNA, this radical is formed via the deprotonation of the guanine radical cation [49] or directly via proton-coupled electron transfer from guanine to an appropriate electron acceptor [10]. The interactions of $G(\text{-}H)^{\bullet}$ with other reactive species have long been of interest. Electron paramagnetic resonance studies performed in neutral aqueous solutions at room temperature have shown that, in the absence of specific reactive molecules, the lifetime of the $G(\text{-}H)^{\bullet}$ radicals in calf thymus DNA is as long as ~5 s [61]. Our own laser flash photolysis studies have shown that the lifetimes of $G(\text{-}H)^{\bullet}$ radicals generated by the oxidation of G residues in 16-mer double-stranded oligonucleotides by carbonate radical anions are about 0.3 s, even in air-equilibrated neutral aqueous solutions [62]. Thus, the $G(\text{-}H)^{\bullet}$ radicals embedded in DNA do not react to any observable extent with molecular oxygen. It is therefore of interest to explore the fates of these long-lived $G(\text{-}H)^{\bullet}$ radicals in DNA in the presence of selected reactive species that rapidly combine with this radical, thus shortening its lifetime. Recently we explored the possibility that the superoxide radical anion ($O_2^{\bullet-}$) derived from trapping hydrated electrons can rapidly react with the $G(\text{-}H)^{\bullet}$ radicals in DNA via a bimolecular combination reaction [11] as shown in Figure 8.5.

The superoxide radical is an important biological intermediate that is formed in living cells [63–65] and also in ionization reactions that generate electrons that are then rapidly captured by dissolved oxygen in aqueous solutions (e.g., [66]). In living systems, superoxide radicals, the products of normal metabolic activity, are

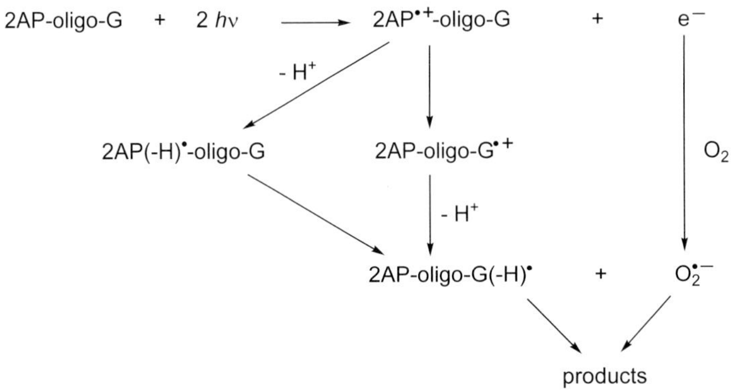

Fig. 8.5 Schematic representation of the photochemical generation of guanine radicals and superoxide radicals via the two-photon photoionization of 2-aminopurine and the subsequent formation of guanine oxidation products in DNA in aqueous solutions.

rapidly deactivated by superoxide dismutases that catalyze the conversion of $O_2^{\bullet -}$ to the less reactive H_2O_2 and O_2 [63–65]. The pioneering work of Fridovich [67], Fielden [68], and their coworkers, based on the pulse radiolysis technique, showed that the catalytic reaction of SOD with $O_2^{\bullet -}$ occurs with nearly diffusion-controlled rates. However, in tissues subjected to chronic infection and inflammation, the consumption of molecular oxygen is enhanced and $O_2^{\bullet -}$ radicals are overproduced [69, 70]. Hence, whenever G(-H)$^\bullet$ radicals in DNA are formed by oxidative one-electron transfer mechanisms followed by deprotonation [71, 72], the reactive combination of G(-H)$^\bullet$ with $O_2^{\bullet -}$ should be feasible in vivo, especially under conditions of oxidative stress.

8.4.1
Spectroscopic Monitoring of the Bimolecular G(-H)$^\bullet$ and $O_2^{\bullet -}$ Combination Reaction

The experiments were performed with the oligonucleotide 5′-d(CC[2AP]TCGCTACC) in either the single-stranded or the double-stranded form [11]. In this oligonucleotide strand, the single G and 2AP residues are separated by two other DNA bases. In 5 mM phosphate buffer solution (pH 7.5) containing 100 mM NaCl, the duplex obtained by annealing the 2AP-modified strand and its natural complementary strand with T opposite 2AP exhibits a single, well-defined cooperative melting curve with a melting point T_m of $50.8 \pm 0.7\,^\circ$C and a hyperchromicity of 25%.

Photoexcitation of the 5′-d(CC[2AP]TCGCTACC) sequence and 5′-d(CC[2AP]TCGCTACC) • 5-d(GGTAGCGATGG) duplex in oxygen-saturated buffer solutions (pH 7.5) with intense 308-nm excimer laser pulses induces selective ionization of the 2AP residues (Figure 8.5). The laser flash photolysis experiments showed that the hydrated electrons are scavenged by molecular oxygen, and thus (1) the formation of superoxide radicals and (2) the oxidation of guanine residues by 2AP radicals are complete within ~100 μs, in agreement with previous observations [6–8]. Therefore, the transient absorption spectra recorded on a millisecond time scale resemble the spectra of guanine neutral radicals (Figure 8.6 a). The G(-H)$^•$ radicals can be identified by the appearance of the characteristic narrow absorption band near 315 nm [6–8].

The gradual decay of the G(-H)$^•$ radicals can be characterized by the change in absorbance at a particular wavelength as a function of time (Figure 8.6 b). This decay is attributed to the combination reaction of the G(-H)$^•$ and $O_2^{•-}$ radicals (Table 8.2, reaction 1). The DNA secondary structure exerts a negligible effect on the radical-radical combination rates since k_1 for the duplex is close to that for the

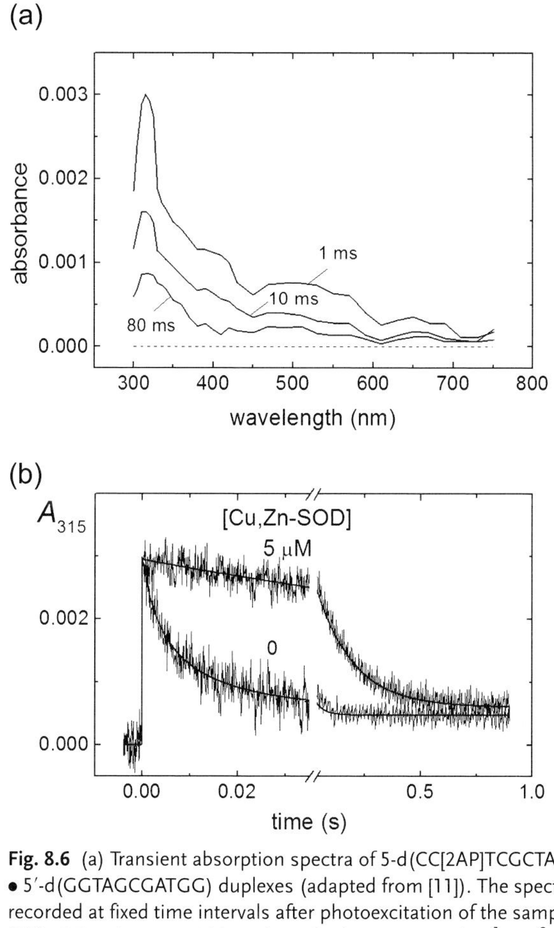

Fig. 8.6 (a) Transient absorption spectra of 5-d(CC[2AP]TCGCTACC) • 5′-d(GGTAGCGATGG) duplexes (adapted from [11]). The spectra were recorded at fixed time intervals after photoexcitation of the samples (100 μM) with actinic 308-nm laser flashes (60 mJ pulse^{-1} cm^{-2}) in 5 mM phosphate buffer solution (pH 7.5) containing 100 mM NaCl and saturated with oxygen at 1 atm ([O_2] ~1.3 mM). (b) Effect of Cu,Zn superoxide dismutase on the kinetics of the combination reaction of the $O_2^{•-}$ and G(-H)$^•$ radicals monitored at 315 nm.

Table 8.2 Kinetic parameters of the bimolecular combination reactions of the G(-H)$^\bullet$ and $O_2^{\bullet-}$ radicals in DNA [11].

No	Reaction	k_n (M^{-1} s^{-1})
1 [a)]	5′-CC[2AP]TC[G(-H)$^\bullet$]CTACC $+ O_2^{\bullet-} \rightarrow$ products	$(4.7 \pm 0.8) \times 10^{8}$ [b)]
	5′-CC[2AP]TC[G(-H)$^\bullet$]CTACC 3′-GG---T--AG----C----GATGG $+ O_2^{\bullet-} \rightarrow$ products	$(4.7 \pm 1.0) \times 10^{8}$
2	$O_2^{\bullet-} + O_2^{\bullet-} + 2H^+ \rightarrow H_2O_2 + O_2$	5.2×10^5 [76]
3	$O_2^{\bullet-}$ + Cu,Zn-SOD $\rightarrow$ products	$(2.3 \pm 0.3) \times 10^9$
4	dG(-H)$^\bullet$ $+ O_2^{\bullet-} \rightarrow$ products	$(3–4) \times 10^9$ [73, 74]
5	dGMP(-H)$^\bullet$ $+ O_2^{\bullet-} \rightarrow$ products	1.3×10^9 [75]
6	$e^- + O_2 \rightarrow O_2^{\bullet-}$	1.9×10^{10} [66]

a) The rate constants were measured in 5 mM phosphate buffer solutions (pH 7.5) containing 100 mM NaCl and saturated with oxygen at 1 atm.
b) The uncertainties are given as standard errors of the best least-squares fits of the appropriate kinetic equations to the transient absorption profiles of the G(-H)$^\bullet$ decay curves recorded at 315 nm.

single strand (Table 8.2). In comparison, the bimolecular rate constants (k_4 and k_5) for the combination of $O_2^{\bullet-}$ and the free base radicals dG(-H)$^\bullet$ (unpublished results cited in [73, 74]) or dGMP(-H)$^\bullet$ [75], where dGMP is 2'-deoxyguanosine 5′-monophosphate, are one order of magnitude greater (Table 8.2). At pH 7.5, the disproportionation reaction of superoxide radicals (reaction 2), is too slow to compete with the combination of the G(-H)$^\bullet$ and $O_2^{\bullet-}$ radicals (reaction 1) [76].

Because the strong absorbance of DNA below 300 nm does not allow for the direct observation of superoxide radicals with an absorption band near 250 nm [77], we utilized other approaches to verify that reaction 1 (Table 8.2) does occur. To obtain support for this conclusion, the G(-H)$^\bullet$ decay kinetics were recorded in the presence of Cu,Zn superoxide dismutase (Cu,Zn-SOD), which is known to react rapidly with $O_2^{\bullet-}$ [67, 68]. Addition of Cu,Zn-SOD at micromolar concentrations exerts a pronounced effect on the kinetics of the G(-H)$^\bullet$ decay (Figure 8.6B). At a 5-μM concentration of Cu,Zn-SOD, the lifetimes of the G(-H)$^\bullet$ radicals are about 0.6 s (single strand) and 0.2 s (duplex), whereas in the absence of Cu,Zn-SOD these lifetimes are ~4 ms and 7 ms, respectively.

The factors that govern the lifetimes of the G(-H)$^\bullet$ radicals in the presence of micromolar concentrations of Cu,Zn-SOD when the effects of $O_2^{\bullet-}$ radicals are minimized are unclear. One of the possible pathways of the decay of G(-H)$^\bullet$ is the reaction of G(-H)$^\bullet$ with molecular oxygen [78]. However, a decrease in oxygen concentration from 1.3 mM in oxygen-saturated solutions to 0.27 mM in air-saturated solutions does not exert any measurable effect on the lifetimes of the G(-H)$^\bullet$ radicals in single- and double-stranded oligonucleotides. The results of these experiments provide an upper limit of $<10^3\ M^{-1}s^{-1}$ for the rate constant of this putative bimolecular reaction, O_2 + G(-H)$^\bullet$ in an oligonucleotide. This result is consistent with the upper limit of $\leq 10^2\ M^{-1}s^{-1}$ for this reaction determined at the nucleoside level [79]. Other pathways for the decay of G(-H)$^\bullet$ radicals may involve unspecified reactions with side products of the laser flash photolysis and nucleophilic reactions with solvent.

8.4.2 Imidazolone Is a Major End Product of G(-H)$^\bullet$ and $O_2^{\bullet-}$ Radical Addition

Reversed-phase HPLC–MALDI-TOF analysis (Figure 8.7 a, b) showed that the major end product of G(-H)$^\bullet$ and $O_2^{\bullet-}$ radical addition in both single- and double-stranded oligonucleotides is the 2,5-diamino-4*H*-imidazolone (Iz) lesion [11]. The latter is slowly hydrolyzed in neutral aqueous solutions to an oxazolone (Z) derivative (Figure 8.7 c) detected by a post-fluorescence HPLC assay [80]. The 8-oxoGua lesions detected in the form of 8-oxodGuo by the HPLC-electrochemical method [81] are formed only in minor quantities (Figure 8.7).

Formation of the end products via the combination of two radicals suggests formation of a covalent bond between these radicals. The G(-H)$^\bullet$ radicals are usually considered to be O-centered radicals with the unpaired electron positioned on the O6 atom [61], which explains the low reactivity of this radical with molecular oxygen [79]. However, formation of the end products can occur via the addition of $O_2^{\bullet-}$ to the C5 and C8 positions of the G(-H)$^\bullet$ radicals. For instance, addition of $^\bullet NO_2$ radicals to the C5 position leads to an unstable adduct that spontaneously collapses to the stable 5-guanidino-4-nitroimidazole lesion, whereas addition of an $^\bullet NO_2$ radical to the C8 position results in the formation of the 8-nitroguanine lesion [82, 83].

Our experiments showed that the combination of G(-H)$^\bullet$ and $O_2^{\bullet-}$ radicals in both single- and doubled-stranded DNA ultimately results in the formation of imidazolone lesions as the major products (Figure 8.7). Cadet and coworkers proposed that the Iz lesions can be formed via the 5-HOO-G(-H) hydroperoxide intermediates [78]. Here, formation of 5-HOO-G(-H) can occur by addition of the $O_2^{\bullet-}$ radical to the C5 position of the G(-H)$^\bullet$ radical followed by a rapid protonation of the peroxide anion (Figure 8.8). Typically, hydroperoxides generated in the course of oxidative degradation of guanine bases are very unstable at room temperature [84]. According to the mechanism proposed by Cadet and coworkers [78], the cleavage of 5-HOO-G(-H) occurs via the opening of the pyrimidine ring at the C5-C6 bond in 5-HOO-G(-H), leading to an unstable intermediate that is easily hydrated at the 7,8–C=N double bond (Figure 8.8). Ring-chain tautomerization of the carbinolamine results in the opening of the imidazole ring, with a subsequent intramolecular cyclization of the guanidine residue resulting in the imidazolone lesion.

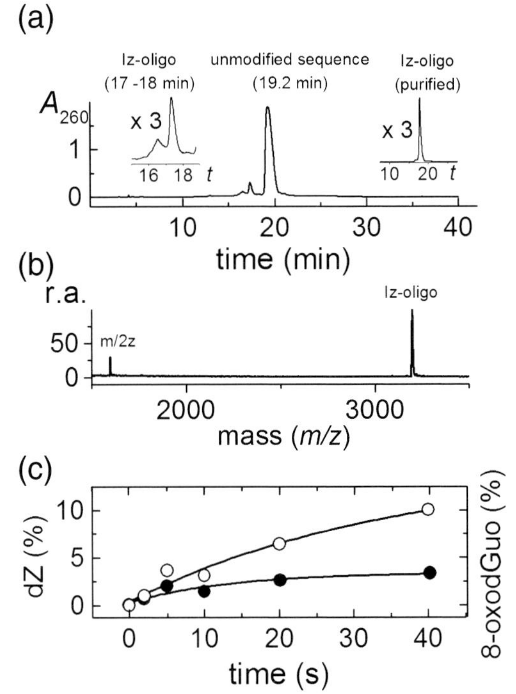

Fig. 8.7 Identification of the guanine oxidation products derived from the combination of $O_2^{\bullet -}$ and G(-H)$^\bullet$ radicals in DNA (adapted from [11]). (a) Reversed-phase HPLC elution profile of the oxidation products. The 5′-d(CC[2AP]TCGCTACC) sequences (100 μM) in 5 mM phosphate buffer solution (pH 7.5) containing 100 mM NaCl and saturated with oxygen at 1 atm ([O_2] ~1.3 mM) were excited by a 10-s train of 308-nm XeCl excimer laser pulses (60 mJ pulse^{-1} cm^{-2}, 10 pulses per second). HPLC elution conditions (detection of products at 260 nm): 10–20% linear gradient of acetonitrile in 50 mM triethylammonium acetate (pH 7) for 60 min at a flow rate of 1 mL min^{-1}. The unmodified sequence 5′-d(CC[2AP]TCGCTACC) eluted at 19.2 min and the damaged oligonucleotides with single imidazolone lesions eluted at 17.4 min. The insert on the left represents an amplified portion of the elution profile, while the insert on the right depicts the elution profile of the purified oligonucleotide containing a single Iz lesion first eluting at 17.4 min, after subsequent cycles of purification. (b) MALDI-TOF negative ion spectrum of the Iz adduct. The 5′-d(CC[2AP]TC[Iz]CTACC adduct was purified by a second HPLC separation. (c) Time dependence of yields of the oxazolone lesions, Z, and 8-oxoGua lesions in 5′-d(CC[2AP]TCGCTACC) duplexes. The oligonucleotide samples (100 μM) were excited under the same conditions as in A. The Z lesions were determined by a post-fluorescence HPLC method [80]. The quantities of dG and 8-oxodGuo were determined by the HPLC-EC method [81] after enzymatic digestion of the irradiated samples. Note that the 8-oxodGuo (right-hand scale) is formed in much lower quantities than dZ.

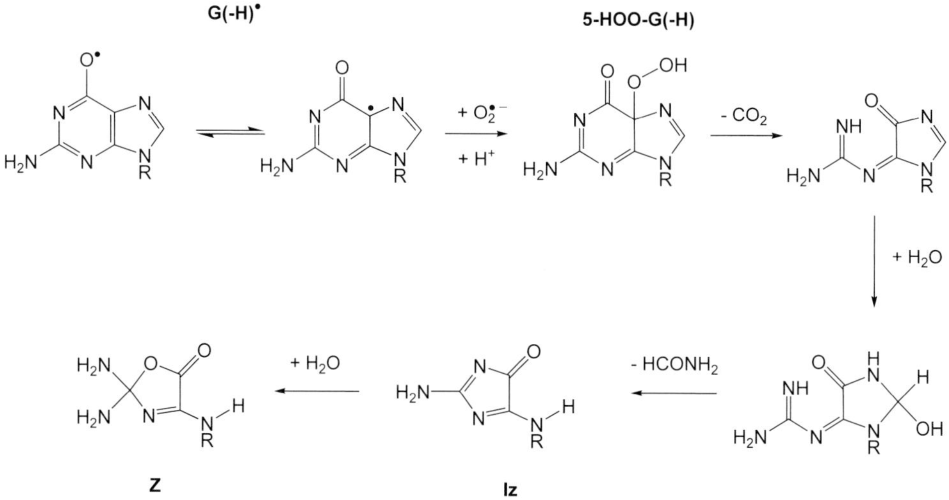

Fig. 8.8 Proposed mechanism of formation of imidazolone adducts via combination of $O_2^{\bullet-}$ and G(-H)$^\bullet$ radicals in DNA.

8.4.3
Minor Pathways Leading to 8-Oxogua Lesions

The chemical reactions that follow the bimolecular combination reaction of the G(-H)$^\bullet$ and $O_2^{\bullet-}$ radicals result in only minor extents of formation of the 8-oxoGua lesion in comparison with the imidazolone lesion (Figure 8.7 c). The low concentrations of 8-oxoGua lesions indicate either a low reaction yield relative to the yield of Iz or a further oxidation of the 8-oxoGua adduct after it has been formed (e.g., [85]). Cadet and coworkers [86] showed that the fast oxidation of 8-oxodGuo by dG(-H)$^\bullet$ can be responsible for the extremely low yields of 8-oxodGuo in the oxidation of dG by excited riboflavin, a typical type I photodynamic agent. In our experiments however, the only guanines present are incorporated into oligonucleotides. Thus, reactions analogous to those observed by Cadet et al. [86] would require a close approach between two negatively charged oligonucleotides, one bearing an 8-oxoGua residue and the other a G(-H)$^\bullet$ radical. Our laser flash photolysis experiments demonstrate that in single- and double-stranded oligonucleotides, the rates of interstrand oxidation of 8-oxoGua by G(-H)$^\bullet$ radicals are two orders of magnitude smaller than the analogous rates for the free nucleosides (Table 8.2). These considerations suggest that, since there are no oxidants present besides G(-H)$^\bullet$ after the termination of the laser irradiation, the oxidation of 8-oxoGua after its formation is unlikely and that the low yields (on the order of ~1%, Figure 8.7 c) are due to low efficiencies of generation of 8-oxoGua.

The experiments in $H_2{}^{18}O$ in the presence of $^{16}O_2$ showed that the 8-oxodGuo formed in the oligonucleotides contains mostly the ^{16}O-isotope, i.e., formation of 8-oxodGuo does not involve addition of $H_2{}^{18}O$ molecules [11]. Molecular oxygen or reactive oxygen species such as superoxide are unique sources of the ^{16}O isotope in the system. Since O_2 does not show any observable reactivity toward G(-H)$^{\bullet}$ radicals, we propose that the formation of the 8-oxoGua lesions involves addition of $^{16}O_2^{\bullet-}$ to the C8 position of the G(-H)$^{\bullet}$ radical. Further reactions of the hydroperoxide 8-HOO-G(-H) (e.g., two-electron reduction) can lead to 8-oxoGua in which the O atom originated from $^{16}O_2$. Decreasing $^{16}O_2$ concentrations by purging the solutions with argon favors the formation of 8-oxoGua with ^{18}O isotope originating from $H_2{}^{18}O$ [11]. This is an indication of the hydration of guanine radicals by $H_2{}^{18}O$ followed by the formation of 8-HO-G$^{\bullet}$ radicals. Pulse radiolysis experiments showed that 8-HO-dG$^{\bullet}$ radicals generated by the addition of hydroxyl radicals to dG are strong reductants that are easily oxidized by weak oxidants such as methyl viologen, $Fe(CN)_6^{3-}$, oxygen [74], and benzoquinone [87], resulting potentially in the formation of 8-oxodGuo. Thus, according to this mechanism, the O atom in 8-oxodGuo originates from $H_2{}^{18}O$. Cadet and coworkers showed that 8-oxodGuo derived from the oxidation of calf thymus DNA by excited riboflavin, a typical type I photodynamic agent, contains O atoms from $H_2{}^{18}O$ [88]. In typical experiments with oxidation of calf thymus DNA by type I photosensitizers such as riboflavin, benzophenone, and menadione in the presence of oxygen, 8-oxoGua and Z lesions form in similar yields [89]. For instance, in the case of riboflavin, these yields were 119 ± 8 of 8-oxoGua and 74 ± 8 Z lesions per 10^6 DNA bases per min. The transient concentrations of the radical species (and $O_2^{\bullet-}$) in the experiments under conditions of continuous excitation should be orders of magnitude lower than in our laser flash photolysis experiments, and thus the hydration of guanine radicals can compete with their combination reactions with superoxide radicals.

8.5 Concluding Remarks

The selective two-photon ionization of 2-aminopurine residues by intense 308-nm excimer laser excitation provides a novel method of site-selective electron abstraction from double-stranded DNA. The primary products, 2AP radical cations, are very short-lived and transform into more stable 2AP neutral radicals that, like the radical cation precursors, are also strong one-electron oxidants. Transient absorption measurements within the nanosecond to millisecond time domains revealed that the kinetics of guanine oxidation at a distance is highly heterogeneous in character. The heterogeneous character of guanine radical formation is associated with the existence of two oxidants that have very different lifetimes in DNA: the 2AP radical cation and the 2AP neutral radical. Spectroscopic kinetic measurements showed that guanine oxidation at a distance in DNA duplexes with adenine-bridging bases between the guanine electron donor and 2AP radical acceptor on the same strand is significantly faster than in the case of bridging thymidines. Using laser and pulse radiolysis kinetic spectroscopy techniques, similar observations have been reported by Lewis, Wasielewski, Majima, and their coworkers [33–36].

The solvent kinetic isotope effects on the rates of guanine oxidation by 2AP neutral radicals indicate that electron transfer reactions from guanine to 2AP radicals occurring at a distance in DNA duplexes are coupled with a deprotonation-protonation mechanism. These proton-coupled electron transfer reactions at a distance generate neutral guanine radicals. Due to the low reactivity toward molecular oxygen and 2′-deoxyribose moieties, guanine radicals are very long-lived (~seconds) in double-stranded DNA.

Trapping reactions of guanine radicals with superoxide radicals derived from scavenging hydrated electrons by molecular oxygen result in the formation of oxidatively modified guanine bases (major product: imidazolone; minor product: 8-oxoGua). Oxygen-18 isotope labeling experiments reveal two pathways of 8-oxo-7,8-dihydroguanine formation, including either addition of $O_2^{\bullet -}$ to the C8 position of G(-H)$^{\bullet}$ (in the presence of dioxygen) or hydration of G(-H)$^{\bullet}$ (in the absence of O_2). Formation of the guanine lesions via combination of guanine and superoxide radicals can be considered as one of the most common pathways of formation of DNA oxidative modifications in photochemical experiments.

Acknowledgments

The research described here was supported by the National Science Foundation (Grant CHE-9700429), by the National Institutes of Health (Grant 5 R01 ES11589), and by a grant from the Kresge Foundation.

References

1 M. E. Nunez, D. B. Hall, J. K. Barton, *Chem. Biol.* **1999**, *6*, 85.
2 G. B. Schuster, *Acc. Chem. Res.* **2000**, *33*, 253.
3 B. Giese, *Annu. Rev. Biochem.* **2002**, *71*, 51.
4 D. B. Hall, R. E. Holmlin, J. K. Barton, *Nature* **1996**, *382*, 731.
5 S. Steenken, S. V. Jovanovic, *J. Am. Chem. Soc.* **1997**, *119*, 617.
6 V. Shafirovich, A. Dourandin, W. Huang, N. P. Luneva, N. E. Geacintov, *J. Phys. Chem. B* **1999**, *103*, 10924.
7 V. Shafirovich, A. Dourandin, W. Huang, N. P. Luneva, N. E. Geacintov, *Phys. Chem. Chem. Phys.* **2000**, *2*, 4399.
8 V. Shafirovich, A. Dourandin, N. E. Geacintov, *J. Phys. Chem. B* **2001**, *105*, 8431.
9 V. Shafirovich, J. Cadet, D. Gasparutto, A. Dourandin, W. Huang, N. E. Geacintov, *J. Phys. Chem. B* **2001**, *105*, 586.
10 V. Shafirovich, N. E. Geacintov, *Top. Curr. Chem.* **2004**, *237*, 129.
11 R. Misiaszek, C. Crean, A. Joffe, N. E. Geacintov, V. Shafirovich, *J. Biol. Chem.* **2004**, *279*, 32106.
12 A. Ronen, *Mutat. Res.* **1980**, *75*, 1.

13 T. M. Nordlund, S. Andersson, L. Nilsson, R. Rigler, A. Gräslund, L. W. McLaughlin, *Biochemistry* **1989**, *28*, 9095.

14 S. W. Law, R. Eritja, M. F. Goodman, K. J. Breslauer, *Biochemistry* **1996**, *35*, 12329.

15 L. C. Sowers, G. V. Fazakerley, R. Eritja, B. E. Kaplan, M. F. Goodman, *Proc. Natl. Acad. Sci. U.S.A.* **1986**, *83*, 5434.

16 P. A. Fagan, C. Fabrega, R. Eritja, M. F. Goodman, D. E. Wemmer, *Biochemistry* **1996**, *35*, 4026.

17 L. C. Sowers, Y. Boulard, G. V. Fazakerley, *Biochemistry* **2000**, *39*, 7613.

18 A. Holmén, B. Nordén, B. Albinsson, *J. Am. Chem. Soc.* **1997**, *119*, 3114.

19 R. Rigler, F. Claesens, in *Structure and Dynamics of RNA* (Eds.: P. H. Van Knippenberg, C. W. Hilbers), Plenum Press, New York, **1986**, pp. 45.

20 C. R. Guest, R. A. Hochstrasser, L. C. Sowers, D. P. Millar, *Biochemistry* **1991**, *30*, 3271.

21 R. A. Hochstrasser, T. E. Carver, L. C. Sowers, D. P. Millar, *Biochemistry* **1994**, *33*, 11971.

22 S. O. Kelley, J. K. Barton, *Science* **1999**, *283*, 375.

23 C. Wan, T. Fiebig, O. Schiemann, J. K. Barton, A. H. Zewail, *Proc. Natl. Acad. Sci. U.S.A.* **2000**, *97*, 14052.

24 O. F. A. Larsen, I. H. M. Van Stokkum, B. Gobets, R. Van Grondelle, H. Van Amerongen, *Biophys. J.* **2001**, *81*, 1115.

25 D. N. Nikogosyan, D. A. Angelov, A. A. Oraevsky, *Photochem. Photobiol.* **1982**, *35*, 627.

26 D. N. Nikogosyan, *Int. J. Radiat. Biol.* **1990**, *57*, 233.

27 L. P. Candeias, S. Steenken, *J. Am. Chem. Soc.* **1989**, *111*, 1094.

28 L. P. Candeias, S. Steenken, *J. Am. Chem. Soc.* **1992**, *114*, 699.

29 V. Shafirovich, A. Dourandin, N. P. Luneva, N. E. Geacintov, *J. Chem. Soc. Perkin Trans. 2* **2000**, 271.

30 M. Gueron, M. Kochoyan, J. L. Leroy, *Nature* **1987**, *328*, 89.

31 M. Kochoyan, J. L. Leroy, M. Gueron, *J. Mol. Biol.* **1987**, *196*, 599.

32 J. L. Leroy, E. Charretier, M. Kochoyan, M. Gueron, *Biochemistry* **1988**, *27*, 8894.

33 F. D. Lewis, X. Zuo, J. Liu, R. T. Hayes, M. R. Wasielewski, *J. Am. Chem. Soc.* **2002**, *124*, 4568.

34 F. D. Lewis, J. Liu, X. Zuo, R. T. Hayes, M. R. Wasielewski, *J. Am. Chem. Soc.* **2003**, *125*, 4850.

35 K. Kawai, T. Takada, S. Tojo, T. Majima, *J Am Chem Soc* **2003**, *125*, 6842.

36 T. Takada, K. Kawai, X. Cai, A. Sugimoto, M. Fujitsuka, T. Majima, *J. Am. Chem. Soc.* **2004**, *126*, 1125.

37 V. Shafirovich, A. Dourandin, N. P. Luneva, N. E. Geacintov, *J. Phys. Chem. B* **2000**, *104*, 137.

38 K. Kawai, T. Takada, S. Tojo, N. Ichinose, T. Majima, *J. Am. Chem. Soc.* **2001**, *123*, 12688.

39 F. D. Lewis, X. Liu, J. Liu, R. T. Hayes, M. R. Wasielewski, *J. Am. Chem. Soc.* **2000**, *122*, 12037.

40 O. I. Kovalsky, I. G. Panyutin, E. I. Budowsky, *Photochem. Photobiol.* **1990**, *52*, 509.
41 K. Ito, S. Inoue, K. Yamamoto, S. Kawanishi, *J. Biol. Chem.* **1993**, *268*, 13221.
42 I. Saito, M. Takayama, H. Sugiyama, K. Nakatani, A. Tsuchida, M. Yamamoto, *J. Am. Chem. Soc.* **1995**, *117*, 6406.
43 J. G. Muller, R. P. Hickerson, R. J. Perez, C. J. Burrows, *J. Am. Chem. Soc.* **1997**, *119*, 1501.
44 I. Saito, T. Nakamura, K. Nakatani, Y. Yoshioka, K. Yamaguchi, H. Sugiyama, *J. Am. Chem. Soc.* **1998**, *120*, 12686.
45 Y. Yoshioka, Y. Kitagawa, Y. Takano, K. Yamaguchi, T. Nakamura, I. Saito, *J. Am. Chem. Soc.* **1999**, *121*, 8712.
46 R. P. Hickerson, F. Prat, J. G. Muller, C. S. Foote, C. J. Burrows, *J. Am. Chem. Soc.* **1999**, *121*, 9423.
47 K. Nakatani, C. Dohno, I. Saito, *J. Am. Chem. Soc.* **2000**, *122*, 5893.
48 H. Decornez, S. Hammes-Schiffer, *J. Phys. Chem. A* **2000**, *104*, 9370.
49 K. Kobayashi, S. Tagawa, *J. Am. Chem. Soc.* **2003**, *125*, 10213.
50 V. A. Kuzmin, A. Dourandin, V. Shafirovich, N. E. Geacintov, *Phys. Chem. Chem. Phys.* **2000**, *2*, 1531.
51 D. O'Connor, V. Y. Shafirovich, N. E. Geacintov, *J. Phys. Chem.* **1994**, *98*, 9831.
52 V. Y. Shafirovich, S. H. Courtney, N. Ya, N. E. Geacintov, *J. Am. Chem. Soc.* **1995**, *117*, 4920.
53 I. Ortmans, B. Elias, J. M. Kelly, C. Moucheron, A. Kirsch-DeMesmaeker, *Dalton Trans* **2004**, 668.
54 S. C. Weatherly, I. V. Yang, H. H. Thorp, *J. Am. Chem. Soc.* **2001**, *123*, 1236.
55 Z. Cai, X. Li, M. D. Sevilla, *J. Phys. Chem. B* **2002**, *106*, 2755.
56 Y. Razskazovskii, M. Roginskaya, M. D. Sevilla, *Radiat. Res.* **1998**, *149*, 422.
57 J. Taylor, I. Eliezer, M. D. Sevilla, *J. Phys. Chem. B* **2001**, *105*, 1614.
58 K. Kawai, Y. Wata, N. Ichinose, T. Majima, *Angew. Chem. Int. Ed.* **2000**, *39*, 4327.
59 K. Kawai, Y. Wata, M. Hara, S. Tojo, T. Majima, *J. Am. Chem. Soc.* **2002**, *124*, 3586.
60 J. R. Milligan, J. A. Aguilera, O. Hoang, A. Ly, N. Q. Tran, J. F. Ward, *J Am Chem Soc* **2004**, *126*, 1682.
61 K. Hildenbrand, D. Schulte-Frohlinde, *Free Radic. Res. Commun.* **1990**, *11*, 195.
62 V. Shafirovich, A. Dourandin, W. Huang, N. E. Geacintov, *J. Biol. Chem.* **2001**, *276*, 24621.
63 I. Fridovich, *Annu. Rev. Biochem.* **1995**, *64*, 97.
64 I. Fridovich, *J. Biol. Chem.* **1997**, *272*, 18515.
65 J. S. Valentine, P. J. Hart, E. B. Gralla, in *Advances in Experimantal Medicine and Biology, Vol. 448* (Eds.: A. Leone, J. F. B. Mercer), Kluwer Academic Publishers, Norwell, MA, **1999**, pp. 193.

66 G. V. Buxton, C. L. Greenstock, W. P. Helman, A. B. Ross, *J. Phys. Chem. Ref. Data* **1988**, *17*, 513.
67 D. Klug, J. Rabani, I. Fridovich, *J. Biol. Chem.* **1972**, *247*, 4839.
68 G. Rotilio, R. C. Bray, E. M. Fielden, *Biochim. Biophys. Acta* **1972**, *268*, 605.
69 M. B. Grisham, D. Jourd'heuil, D. A. Wink, *Aliment. Pharmacol. Ther.* **2000**, *14 Suppl 1*, 3.
70 B. Halliwell, *Mutat. Res.* **1999**, *443*, 37.
71 S. Steenken, *Chem. Rev.* **1989**, *89*, 503.
72 S. Steenken, *Biol. Chem.* **1997**, *378*, 1293.
73 C. von Sonntag, *Int. J. Radiat. Biol.* **1994**, *66*, 485.
74 L. P. Candeias, S. Steenken, *Chem. Eur. J.* **2000**, *6*, 475.
75 V. Shafirovich, J. Cadet, D. Gasparutto, A. Dourandin, N. E. Geacintov, *Chem. Res. Toxicol.* **2001**, *14*, 233.
76 B. H. J. Bielski, D. E. Cabelli, R. L. Arudi, A. B. Ross, *J. Phys. Chem. Ref. Data* **1985**, *14*, 1041.
77 B. H. J. Bielski, *Photochem. Photobiol.* **1978**, *28*, 645.
78 J. Cadet, M. Berger, G. W. Buchko, P. C. Joshi, S. Raoul, J.-L. Ravanat, *J. Am. Chem. Soc.* **1994**, *116*, 7403.
79 M. Al-Sheikhly, *Radiat. Phys. Chem.* **1994**, *44*, 297.
80 J. Ravanat, M. Berger, F. Benard, R. Langlois, R. Ouellet, J. E. van Lier, J. Cadet, *Photochem. Photobiol.* **1992**, *55*, 809.
81 M. Berger, C. Anselmino, J.-F. Mouret, J. Cadet, *J. Liq. Chromatogr.* **1990**, *13*, 929.
82 V. Shafirovich, S. Mock, A. Kolbanovskiy, N. E. Geacintov, *Chem. Res. Toxicol.* **2002**, *15*, 591.
83 A. Joffe, S. Mock, B. H. Yun, A. Kolbanovskiy, N. E. Geacintov, V. Shafirovich, *Chem Res Toxicol* **2003**, *16*, 966.
84 P. Kang, C. S. Foote, *J. Am. Chem. Soc.* **2002**, *124*, 4865.
85 A. Joffe, N. E. Geacintov, V. Shafirovich, *Chem. Res. Toxicol.* **2003**, *16*, 1528.
86 J. L. Ravanat, C. Saint-Pierre, J. Cadet, *J. Am. Chem. Soc.* **2003**, *125*, 2030.
87 M. G. Simic, S. V. Jovanovic, in *Mechanisms of DNA damage and repair: implications for carcinogenesis and risk assessment, Vol. 38* (Eds.: M. G. Simic, L. Grossman, A. C. Upton), Plenum Press, New York, **1986**, pp. 39.
88 H. Kasai, Z. Yamaizumi, M. Berger, J. Cadet, *J. Am. Chem. Soc.* **1992**, *114*, 9692.
89 T. Douki, J. Cadet, *Int. J. Radiat. Biol.* **1999**, *75*, 571.

9
Electron Transfer and Structural Dynamics in DNA

Hans-Achim Wagenknecht and Torsten Fiebig

9.1
Introduction

Electron transfer (ET) reactions represent a widespread and exciting research field in chemistry as well as in biology. In particular, the biomacromolecule DNA as a unique medium for ET has attracted a considerable amount of interest. The efforts of our research groups are directed towards two major goals: (1) probing DNA dynamics on the ultrafast time scale and (2) investigating electron injection into DNA and transfer through DNA. Both subjects are strongly interconnected. Recent experimental [1–4] and theoretical work [5–11] has emphasized that the electronic properties of DNA are dependent upon its geometrical structure and thus on time-dependent structural fluctuations and motions. In this chapter we review our own efforts to assess the interplay between structural fluctuations – which provide the basis for conformational gating [8] – and ET dynamics in DNA.

With respect to the short excited-state lifetimes of the natural DNA bases [12], it becomes necessary to modify oligonucleotides with suitable chromophores as "chemical probes" in order to photoinitiate electron injection processes into the DNA [13]. It is important to point out that a profound understanding of the interaction between the charge donor and the base stack of the DNA helix is crucial for the clear assignment and interpretation of the resulting ET processes.

Great synthetic efforts in recent years have led to a large number of functionalized oligonucleotides with more or less well-defined D-A distances and geometries [13–15]. Although supporting structural data from NMR or X-ray crystallography are only scarcely available, these molecular systems are inherently superior to their early predecessors with noncovalently attached chromophores [14]. The systems presented here contain chromophores that are attached covalently and, more importantly, site-specifically to oligonucleotides. Depending on the type of chromophore, we chose either to modify DNA bases by chromophores that point into the major groove or to substitute single natural bases by artificial ones (Figure 9.1). Structures of type B include chromophore interactions with five bases located in the direct vicinity of the chromophore. Molecular structure models, how-

Charge Transfer in DNA: From Mechanism to Application. Edited by H.-A. Wagenknecht

ISBN: 3-527-31085-1

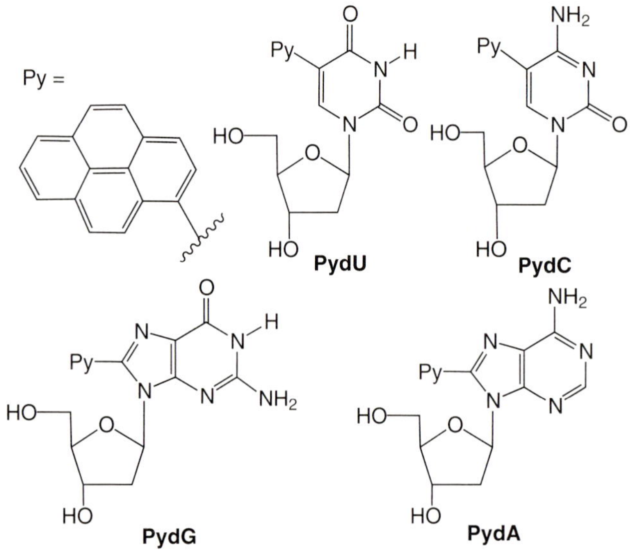

Fig. 9.2 Pyrene-modified nucleosides as models for charge transfer in DNA.

thetic trick, we can decide whether an intramolecular electron transfer or a hole transfer occurs (Figure 9.2). In the case of **PydU** and 5-(pyren-1-yl)-2′-deoxycytidine (**PydC**), excitation of the pyrene moiety leads to an intramolecular electron transfer yielding the corresponding pyrimidine radical anions ($Py^{\bullet+}dU^{\bullet-}$ and $Py^{\bullet+}dC^{\bullet-}$). According to our studies, both uracil and cytosine are significantly weaker electron acceptors than expected based upon redox potentials. Combining the potentials $E(Py^{\bullet+}/Py^{*}) = -1.8$ V (vs. NHE) [30] and $E(dU/dU^{\bullet-}) = E(C/C^{\bullet-}) = -1.1$ V [31], the driving force ΔG for the ET process in **PydU** or **PydC** could be maximally –0.6 eV (using $E_{00} = 3.25$ eV for Py^{*} [30]). However, our studies revealed a driving force $\Delta G \approx 0$ eV, which requires the potential $E(dU/dU^{\bullet-})/E(C/C^{\bullet-})$ to be ~ -1.8 V [32, 33]. In this context, the measured value $E(dU/dU^{\bullet-}) = -1.1$ V provided by Steenken et al. [31] is difficult to understand and could reflect the result of a proton-coupled ET [33]. Thus, it is likely, that the –1.1 V potential corresponds to $E(dU/dU(H))^{\bullet}$. In contrast, excitation of 8-(pyren-1-yl)-2′-deoxyguanosine (**PydG**) results in formation of the guanine radical cation as the product of an intramolecular HT ($Py^{\bullet-}$ $dG^{\bullet+}$), which formally represents an electron transfer in the opposite direction in **PydU** and **PydC**. Based on $E_{00} = 3.25$ eV for Py^{*} [30], and using $E^{0}(Py/Py^{\bullet-}) = -1.9$ V [30] and $E^{0}(G^{\bullet+}/G) = 1.3$ V [34], the driving force ΔG of this process is around -0.05 eV. A prediction of the charge transfer direction in 8-(pyren-1-yl)-2′-deoxyadenosine (**PydA**) is difficult. Based on the redox potentials of adenine [34, 35], both oxidation and reduction of the adenine moiety in **PydA** seem to be slightly unfavorable.

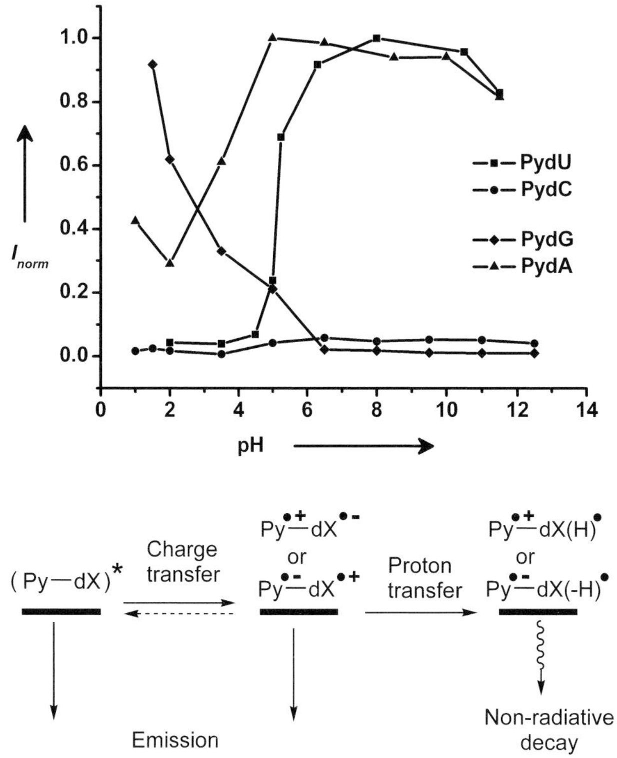

Fig. 9.3 pH dependence of the fluorescence intensity of **PydU**, **PydC**, **PydG**, and **PydA** in water at equal optical density (excitation at 340 nm).

9.2.2
pH-dependent Fluorescence Quenching

The fluorescence properties of the synthesized pyrene-modified nucleosides in water at different pH values reflect information about the acid-base properties of the generated DNA base radicals [36]. The pH dependence of the emission of **PydU** shows a typical sigmoidal curve representing a pK_a value of ~5 for the protonated biradical $Py^{\bullet+}$ $dU(H)^{\bullet}$ (Figure 9.3). In contrast, **PydC** exhibits fluorescence quenching over almost the entire pH range, indicating that the pK_a value for the protonated biradical $Py^{\bullet+}$ $dC(H)^{\bullet}$ must be larger than 12. Remarkably, these results are similar to the published pK_a values by Steenken et al. [37]. It is important to point out that the unprotonated charge-separated species $Py^{\bullet+}$ $dC^{\bullet-}$ cannot be observed in water. The remarkable difference in the basicities of the generated pyrimidine radical anions is significant for the mechanism of electron migration. Based on the reduction potentials, it was proposed that both $C^{\bullet-}$ and $T^{\bullet-}$ (structurally very similar to $dU^{\bullet-}$) could act as potential intermediates in electron migration through DNA. Our results suggest that proton transfer does not limit electron hopping via A-T base pairs but significantly interferes with electron transport through G-C base pairs.

The steady-state fluorescence of **PydG** in water exhibits an inverted pH dependence to the one observed for **PydU**. These emission vs. pH profiles demonstrate that HT (instead of ET) occurs in **PydG**, generating a cationic nucleoside species ($Py^{\bullet -}-dG^{\bullet +}$) that deprotonates at higher pH values. The question of proton transfer in $G^{\bullet +}C$ base pairs is crucial for the understanding of hole hopping in DNA. Based on our results, the pK_a value of $G^{\bullet +}$ is ~4. The pK_a value of the complementary DNA base C is very similar (4.5) [37]. Hence, there is likely a protonation equilibrium in a one-electron-oxidized $G^{\bullet +}$-C base pair that could interfere with the hole transport and potentially interrupt hole hopping in DNA. In fact, measurements of the kinetic isotope effect of hole transport in DNA performed by Giese et al. provide some evidence for a coupling between hole-hopping and proton-transfer processes [38].

As already mentioned above, in the case of **PydA** a prediction of the type of charge-transfer process (i.e., ET vs. HT) based on redox potentials appears ambiguous. It is remarkable that the emission profile of **PydA** in water at different pH values (Figure 9.3) provides the answer. Comparing the pH-dependent emission of **PydA** with that of **PydU** vs. **PydG**, it becomes clear that a reduction of the A moiety takes place, but the measured pK_a value of ~4 appears to be rather unlikely for the protonated biradical $Py^{\bullet +}$ $dA(H)^{\bullet}$ with respect to the 6-aminopurine substructure as part of **PydA** [37]. Hence, the fluorescence quenching of **PydA** at pH <3 is concluded to be the result of a ground-state protonation.

9.2.3
The Role of Hydrogen Bonding in ET Dynamics

Netzel et al. reported the steady-state fluorescence spectra of **PydU** in two polar organic solvents, the non-protic acetonitrile MeCN and the protic methanol MeOH [24, 25]. Since the fluorescence spectrum in MeOH clearly indicates the presence of an exciplex-like charge-transfer emission (which is far less strongly pronounced in MeCN!) [29], it was concluded that **PydU** undergoes a proton-coupled electron transfer (PCET) in MeOH. However, this interpretation is weakened by the fact that protonation of the uridine radical anion should be energetically highly unfavorable in MeOH.

We have applied femtosecond broadband pump-probe spectroscopy to illuminate the early excited-state dynamics in **PydU** after optical excitation. Our data have confirmed the involvement of the hydrogen-bonding network in the ET process and identified hydrogen-bond dynamics as a key driving force for ET in **PydU**.

Figure 9.4 shows the time-resolved transient absorption spectra of **PydU** in MeOH (Figure 9.4 a) and MeCN (Figure 9.4 b) in the time range between 3 ps and 20 ps after excitation. In addition to the *local* pyrenyl absorption bands (Py^* and $Py^{\bullet +}$) between 470 nm and 520 nm, the spectrum shows a broad and intense band in the near-infrared region (550–700 nm). This band cannot be ascribed to

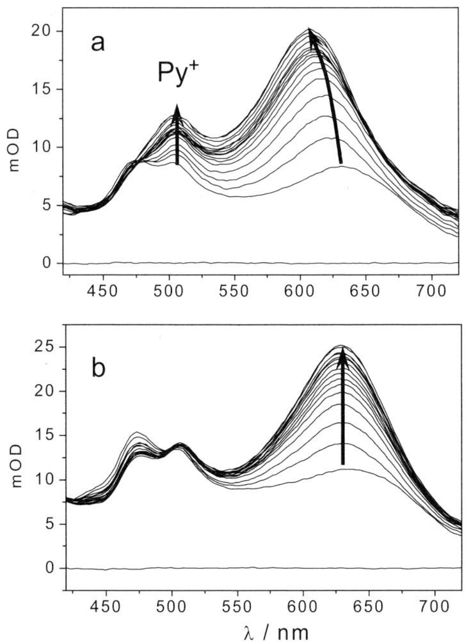

Fig. 9.4. Time-dependent evolution of the pump-probe spectrum of **PydU** in (a) MeOH and (b) MeCN, each between 3 ps and 20 ps after excitation.

either (local) pyrenyl or uridine transitions. It rather corresponds to a contact ion pair (CIP) absorption which is a composite of the two strongly interacting π-systems in $Py^{\bullet+}$ $dU^{\bullet}$ [39, 40]. In contrast to local radical ion bands, the intensity of the CIP band depends on the magnitude of the electronic coupling between the two aromatic subsystems and therefore on the conformation of **PydU** [32]. A rise in the intensity suggests the transition from a twisted to a more planar geometry with enhanced π-electron overlap between Py and dU [41]. The fact that in MeOH the $Py^{\bullet+}$ absorption (505 nm) rises on the same time scale as the CIP band (see Figure 9.4 a) clearly indicates that the ET process is directly coupled to a structural change in the molecule. In other words, while the molecule is moving towards a more planar geometry, more charge is transferred from Py^* to dU. In MeCN, on the other hand, one observes a fast equilibration between Py^* and $Py^{\bullet+}$ on the time scale of a few picoseconds, which is typical for electron D-A systems with small driving forces where the ET is incomplete (Figure 9.4 b). Subsequent to ET the molecule undergoes a structural relaxation similar to that observed in MeOH, indicated by the rise of the CIP band around 630 nm. In contrast to the MeOH data, however, the CIP band in MeCN does not exhibit a blue shift. The absence of the spectral shift in MeCN indicates that the shift is caused by MeOH-specific solvation interactions. These interactions are likely caused by the hydrogen-bond network, particularly the H bonds between MeOH and $dU/dU^{\bullet-}$. As mentioned above, the protonation of $dU^{\bullet-}$ by MeOH should be thermodynamically unfavorable. This interpretation is based on the pK_a value of 6.9 determined by Steenken et al. for the protonated thymidine radical $dT(H)^{\bullet}$, which thus represents a stronger acid than MeOH (pK_a 15.7) [37]. Based on the difference in pK_a values, one would expect a rate constant for $dU^{\bullet-}$ protonation on the order of milliseconds, i.e., several orders of magnitude slower than the lifetime of the CIP state [42]. Instead we proposed a "proton-driven" ET mechanism where hydrogen bonding facilitates the transfer of electronic charge from Py* to dU [43]. While protons will certainly be "adjusted" within the hydrogen-bond network, there should not be a kinetically distinguishable $Py^{\bullet+}dU(H)^{\bullet}$ species.

In earlier studies we were able to induce a PCET mechanism by dramatically changing the pH value in an aqueous solution of **PydU** [32, 36]. However, these pH values are drastically different from physiological conditions and do not match the protic environment within the DNA base stack. The fact that the ET dynamics are so dramatically different in MeOH (as compared to MeCN) provides a clear indication that the hydrogen-bond network participates actively in the transfer dynamics.

9.3
Pyrene as an Electron Injector and a Probe for Base Dynamics in DNA

On the basis of the findings described in the previous section, it can be concluded that **PydU** is an optimally tuned intramolecular ET system with high sensitivity for hydrogen-bond dynamics. As such, the next logical step was to covalently "build" **PydU** into the DNA base stack. A second motivation for **PydU** in DNA was to use it in order to inject electrons from the photoexcited Py* to dU and thus into the base stack. Using the corresponding DNA building block of **PydU** in automated phosphoramidite chemistry, we prepared a set of **PydU**-modified DNA duplexes (Figure 9.5). The duplexes **DNA-1–DNA-4** contain only a single modification, which is the **PydU** chromophore as the photoexcitable electron injector. C and T represent the closest possible electron acceptor for electron hopping and they

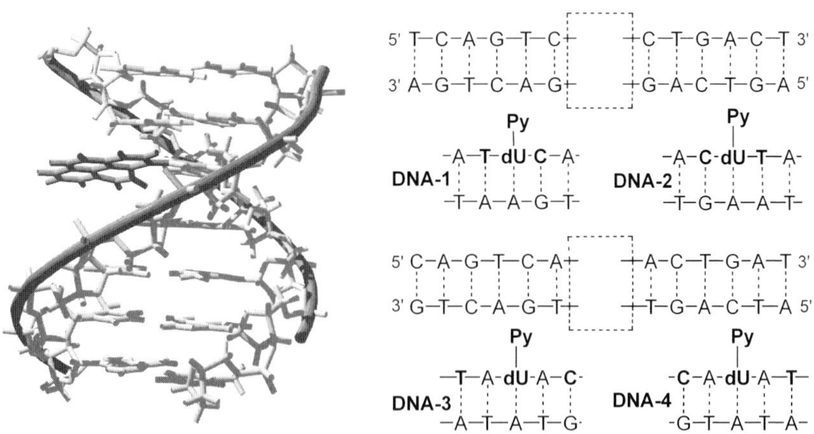

Fig. 9.5 **PydU**-modified duplexes **DNA-1–DNA-4**.

either are placed adjacent to the **PydU** group or are separated from **PydU** by one intervening A-T base pair.

9.3.1
Ultrafast Spectral Shift – Hydrogen-Bond Dynamics in DNA

As for **PydU** in organic solvents, we applied femtosecond broadband pump-probe spectroscopy to explore the early-time dynamics of **PydU** in DNA as surrounding medium. Upon excitation at 350 nm, a pyrene-like excited state (Py-dU)* is formed that undergoes ET to form the CIP state $Py^{\bullet+}dU^{\bullet-}$. Since the latter process formally represents the injection of an electron into the base stack, one can obtain the injection rate from the decay of the transient absorption band of (Py-dU)* at 385 nm, the well-known excited, singlet-state absorption of pyrene [44, 45]. This decay time – which also appears as a rise component in the CIP absorption – is 2–3 ps in **DNA-1–DNA-4**, i.e., about 20 times faster than in MeOH. However, the broad absorption band of the CIP state undergoes an ultrafast blue shift, similar to the one observed in MeOH.

From the time-dependent spectral shift of the CIP band, a peak-shift correlation function (PSCF) can be formulated:

$$C(t) = \frac{\nu(t) - \nu(\infty)}{\nu(0) - \nu(\infty)} \quad (1)$$

where $\nu(t)$ is the maximum of the excited-state absorption spectrum at time t. Equation (1) has been used extensively to characterize solvation dynamics by time-resolved fluorescence Stokes-shift measurements [46, 47]. Here, we chose the analogous treatment of DNA as a solvating medium that actively controls ET dynamics.

Figure 9.6 shows $C(t)$ for **PydU** in MeOH, ACN, and DNA (**DNA-1–DNA-4**). $C(t)$ has been fitted with a biexponential decay function (see Table 9.1). It is remarkable that the longer decay component of the PSCF for the DNA systems is substantially longer than for PydU in MeCN but shorter than for PydU in MeOH. In fact, Figure 9.6 shows that the decay of PSCF at a longer time (>1 ps) falls right between the PSCFs in MeOH and MeCN. This observation and the fact that the overall spectral dynamics of **PydU** are very similar in MeOH and in “DNA” provide strong support for the following thesis: hydrogen bonding and changes in the hydrogen-bonding configurations play an essential role for transferring an electron to uracil, especially if the transfer occurs in the small driving force regime. In DNA, the hydrogen-bonding configuration is mainly dictated by the complementary base. But external molecules such as peptides and proteins that bind to DNA can form hydrogen bonds with bases, thereby transiently (and permanently) alternating their redox properties. Additional experiments on **PydU** derivatives will be needed to further explore the role of hydrogen bonding in charge-transfer processes.

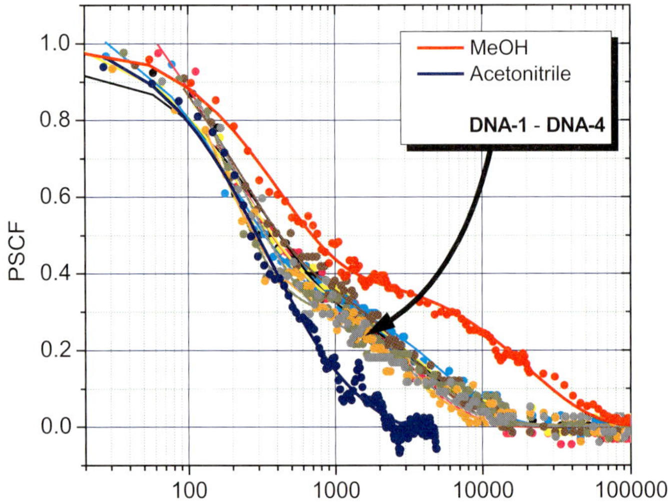

Fig. 9.6 Peak-shift correlation functions (PSCF) of **PydU** in MeOH, MeCN and of **PydU**-modified **DNA-1–DNA-4**.

Table 9.1 Characteristic time constants of the peak-shift correlation function (PSCF, Eq. 1) for **PydU** in MeOH, in MeCN, and in DNA obtained from a least-squares fit using a biexponential decay function. "DNA" refers to averaged results for **DNA1–DNA4**.

	τ_1 (ps)	τ_2 (ps)	A_1 (%)	A_2 (%)
MeOH	0.4	21	40	60
MeCN	0.25	0.9	60	40
"DNA"	~0.2	~4.5	55	45

9.3.2
Probing the ET Products in DNA – Dispersive Kinetics and Strand Cleavage

The pump-probe transients of all four **PydU**-modified duplexes between 440 nm and 700 nm show biexponential decays, with a nanosecond (≥ 3 ns) and a picosecond decay component (Figure 9.7). The latter changes continuously with the probe wavelength from 120 ps to 600 ps (in **DNA-1** between 440 nm and 700 nm). This strong kinetic dispersion in the lifetimes of the CIP states is consistent with multi-conformational substrates in a largely disordered medium [48].

To answer the question of whether subsequent electron shift into the base stack competes with charge recombination in the CIP state, we measured the repopulation dynamics of the **PydU** ground state by monitoring the ground-state bleaching

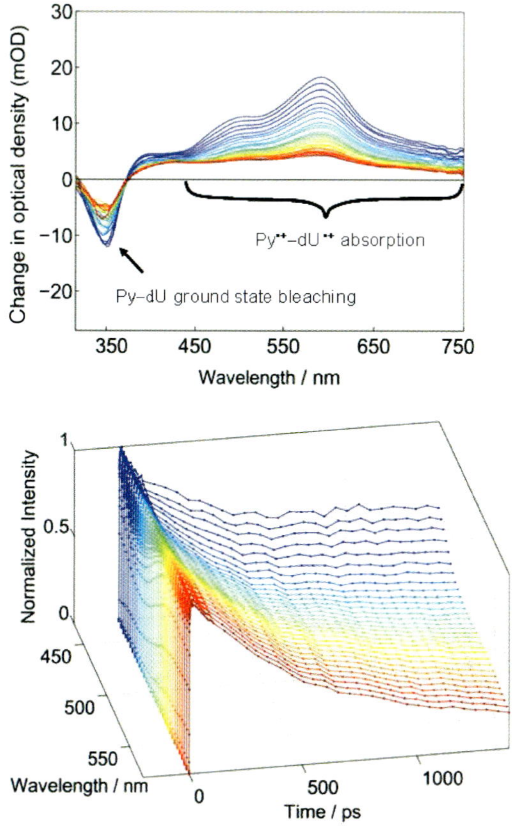

Fig. 9.7 Top: Time-dependent decay of the pump-probe spectra 150 ps (blue) to 1.5 ns (red) after photoexcitation of **DNA-1** Bottom: Normalized $Py^{\bullet+}dU^{\bullet-}$ transient absorption signals between 430 nm and 580 nm.

signal around 350 nm. Interestingly, we did not find the same dispersion of picosecond rate constants as observed for the decays of the CIP state. Instead, the ground state of **DNA-1** is repopulated with a distinct picosecond time constant of 200 ± 50 ps. This observation suggests that only a fraction of CIP ensembles return to the ground state. Thus, the remaining CIP populations are reacting through a different channel to an electronic state other than the ground state. To identify the nature of this state, we probed the product state of this multi-step, DNA-mediated ET process chemically, i.e., via strand cleavage experiments. For this purpose, a second DNA set was synthesized, **DNA-5–DNA-8**, containing 5-bromo-2′-deoxyuridine (**BrdU**) as an additional modification [49].

It is known that **BrdU** undergoes a chemical modification after its one-electron reduction that can be analyzed by piperidine-induced strand cleavage [50, 51]. Hence, the quantification of the strand cleavage yields information about the ET efficiency, as recently shown by Rokita et al. [52, 53]. It is important to point out that based on reduction potentials **BrdU** is not a significantly better electron acceptor and that very similar rates for the decay of the CIP state were measured in **DNA-5** and **DNA-6** (with **BrdU** adjacent to **PydU**) compared to **DNA-1** and **DNA-2** (with C/T adjacent to **PydU**). Hence, **BrdU** represents a *kinetic* electron trap. Strand degradation can be observed during the irradiation of all four **PydU**-modified DNA duplexes. Interestingly, **DNA-5** and **DNA-6**, where **BrdU** is directly adjacent to **PydU**, show much higher cleavage efficiency compared to **DNA-7** and **DNA-8**, where **BrdU** is separated from **PydU** by one intervening A-T base pair (Figure 9.8). Thus, considering the fact that strand degradation represents the chemical result of the DNA-mediated ET process, it is remarkable that just one intervening A-T base pair lowers the ET efficiency between **PydU** and **BrdU** to such an extent. This result indicates that conformational control of ET in DNA becomes more dominant with increasing distances, which is entirely consistent with the observed dispersion of CIP lifetimes.

There are several important conclusions emerging from this study. First, DNA is a disordered medium with a manifold of conformational states exhibiting a wide range of reactivities and rate constants. Second, as expected, the electron injection process in our functionalized DNA sets shows only minor variations due to structural inhomogeneity, because the injection process occurs between the covalently connected Py and dU moiety. Subsequent electron shift into the base stack – as evidenced by the strand cleavage experiments – is much more sensitive to structural parameters and thereby is characterized by a distribution of time constants. Third, our results demonstrate the importance of probing both the early-time events and the product states for obtaining conclusive mechanistic insight. Since DNA-mediated ET is a multi-step process that occurs on various time scales, the measured electron injection rates may not necessarily correlate with the ob-

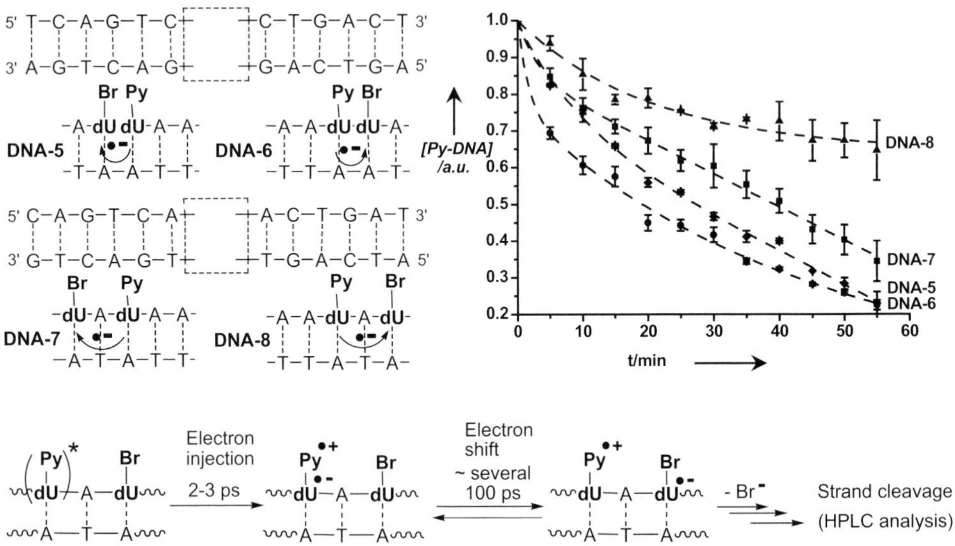

Fig. 9.8 Strand cleavage experiments with **PydU**-modified **DNA-5**–**DNA-8**.

served strand cleavage efficiencies as the chemical result of DNA-mediated ET. Finally, our results indicate that the electron shift occurs on the same time scale (several hundred picoseconds) as charge recombination from the CIP to the ground state. It is reasonable to assume that additional migration steps will be faster since the Coulomb interaction between the electron and $Py^{\bullet +}$ decreases drastically with distance. Our results therefore provide a lower limit for excess electron transfer rates through DNA.

9.4 Reductive Electron Transfer in Phenothiazine-modified DNA

In comparison to pyrene as the electron donor, the reduction potential of phenothiazine (Ptz) in the excited state $E(Ptz^{\bullet +}/Ptz^{*}) = -2.0$ V [54, 55] should be more efficient for the photoreduction of T and C within DNA. Accordingly, we synthesized **PtzdU** and incorporated it into oligonucleotides [56]. The sequences of the duplexes **DNA-9–DNA-14** have been designed in such a way that the **BrdU** group as the electron acceptor is placed two, three, or four base pairs away from the **PtzdU** group as the photoexcitable electron donor. By using this approach, we wanted to elucidate the distance and sequence dependence of DNA-mediated ET by chemical means. For excess electron transfer over long distances, a hopping mechanism involving $C^{\bullet -}$ and $T^{\bullet -}$ as intermediate electron carriers was suggested [57]. Accordingly, the intervening base pairs X-Y of **DNA-9–DNA-14** were chosen to be either T-A or C-G (Figure 9.9).

Indeed, strand degradation can be observed during the irradiation of all **PtzdU**-modified duplexes, but the efficiencies of the strand cleavage show significant differences. Interestingly, the DNA duplexes with the intervening T-A base pairs (**DNA-9**, **DNA-11**, and **DNA-13**) show a significantly higher cleavage efficiency compared to the DNA duplexes with the intervening C-G base pairs (**DNA-10**, **DNA-12**, and **DNA-14**). In fact, the cleavage efficiency of **DNA-13** is comparable to that of **DNA-10**. Thus, considering the fact that strand degradation represents the chemical result of the DNA-mediated ET process, it is remarkable that just one intervening C-G base pair exhibits an ET efficiency similar to that of three intervening T-A base pairs. From these strand cleavage experiments, it becomes clear that in our assay T-A base pairs transport electrons more efficiently than C-G base pairs (Figure 9.10). This implies that $C^{\bullet -}$ is not likely to play a major role as an intermediate electron carrier.

This observation is supported by a number of recent publications. Rokita applied aromatic amines as electron donors together with **BrdU** as electron trap and could show that the ET efficiency significantly depends on the intervening base sequence. The presence of C-G base pairs lowered the ET efficiency significantly [53]. Sevilla employed EPR spectroscopy and showed that proton transfer can slow down excess electron transfer but does not stop it [58]. The latter result is consistent with our spectroscopic studies using **PydU** and **PydC** as models for ET in DNA, as mentioned above. It is very likely that the protonation of $C^{\bullet -}$ by the com-

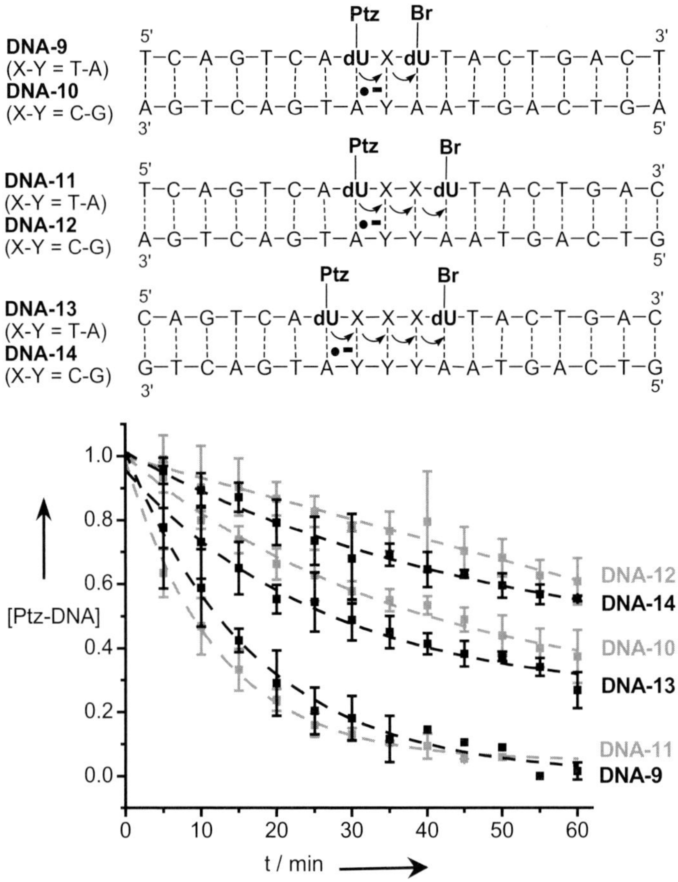

Fig. 9.9 Strand cleavage experiments with **PtzdU**-modified **DNA-9–DNA-14.**

plementary DNA base G or the surrounding water molecules will occur rapidly. Although the hydrogen bond interface can readjust and is microscopically reversible, proton transfer may ultimately terminate electron migration in DNA due to the separation of spin and charge. In conclusion, it is now clear that excess electron transfer via hopping is highly sequence dependent and occurs faster and more efficiently over T-A base pairs than over C-G base pairs.

9.5
Structural Flexibility and Base Dynamics in Py≡dU-Modified DNA

Although an acetylene bridge is a strong mediator for electronic coupling [59–61] between chromophores and dU, the additional separation between Py and dU inhibits a complete electron transfer. The ethynylpyrene chromophore (Py≡) has

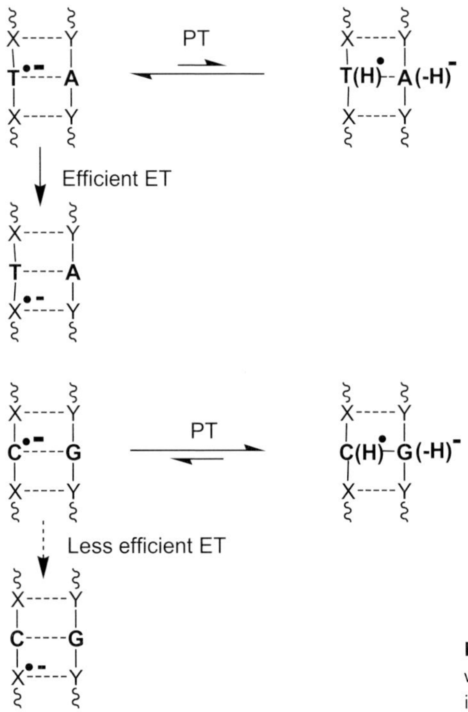

Fig. 9.10 Proton transfer (PT) interferes with ET in C-G base pairs (bottom) but not in T-A base pairs.

been covalently attached to the 5 position of dU, first synthesized by Berlin et al. [62, 63]. Thus, the chromophore replaces the methyl group of naturally occurring thymidines and is expected to be located in the major groove [64, 65]. This assumption is based on similar modifications in which such ethynyl substitutions have little influence on the stability of the resulting modified DNA duplexes and do not perturb the Watson-Crick base-pairing ability [64, 65]. Thus, the ethynylpyrene group extends the conjugated planar system of dU and equips it with optically traceable properties. Note that by using two Py≡ modifications we are able to probe the dynamics of the intervening base pair. This would not be possible (at least not in such a regioselective way) when using the naturally occurring DNA bases.

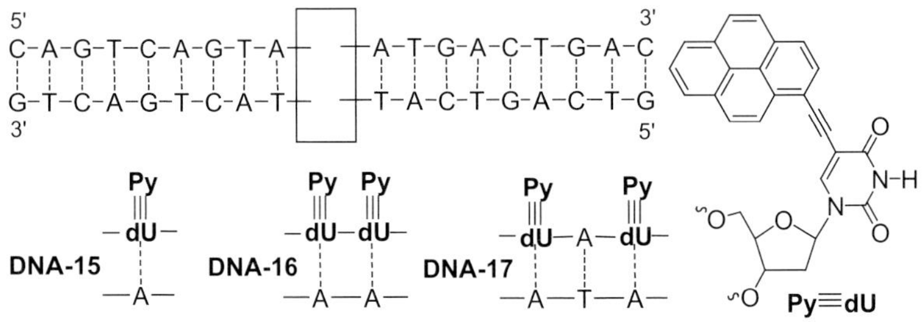

Fig. 9.11 **Py≡dX**-modified **DNA-15**, **DNA-16**, and **DNA-17**.

Three **Py≡dU**-modified DNA duplexes (**DNA-15–DNA-17**, Figure 9.11) have been synthesized by a semiautomated solid-phase strategy [66] and investigated as molecular beacons in DNA hybridization experiments [67].

9.5.1 Femtosecond Broadband Pump-Probe Spectroscopy

Between 450 nm and 750 nm the pump-probe spectra of **DNA-15–DNA-17** are dominated by excited-state absorption (Figure 9.12). In all three **Py≡dU**-modified DNA duplexes, at very early times (within 100 fs after excitation) one observes a broad, asymmetric absorption band with a maximum around 700 nm. This absorption band originates from the excited **Py≡dU** moiety. Due to the strong electronic coupling across the C-C triple bond, **Py≡dU** must be regarded as a single chromophore where the electronic excitation energy is widely delocalized over the entire system; hence, the 700 nm absorption is an interchromophore (IC) band. Although a complete electron transfer from Py* to dU can be ruled out, the delocalized excited-state (Py≡dU)* is likely to have partial charge-transfer character. The onset of relaxation in the excited state is manifested by a spectral blue shift of the IC band.

It is interesting to compare the spectral shifting dynamics of **DNA-15–DNA-17**. Figure 9.13 shows the spectral position of the maximum of the absorption band

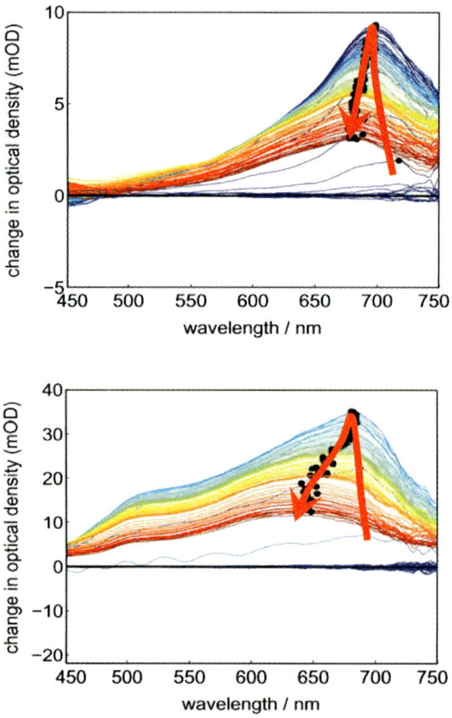

Fig. 9.12 Temporal evolution of the pump-probe spectra of **DNA-15** (top left), **DNA-16** (top right), and **DNA-17** (bottom) after excitation at 364 nm.

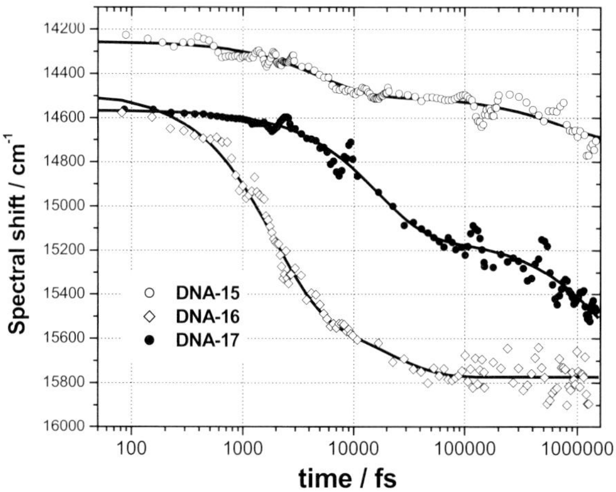

Fig. 9.13 Time-dependent peak shift of the maxima of the transient absorption spectra of **Py**≡**dX**-modified **DNA15–DNA-17.**

evolving in time. All peak-shift functions (PSFs) have been fitted using biexponential functions (Table 9.2).

In **DNA-15**, the IC band undergoes a rapid shift with a time constant of 4.6 ps (70%), followed by a slower shift of 220 ps (30%). However, the total magnitude of the shift is only 440 cm^{-1}. In contrast, in **DNA-16** the spectral shift is 1300 cm^{-1} with a time constant of ~2 ps for 85%. In **DNA-17**, one observes almost the same magnitude of spectral shifting, exhibited, however, on a much longer time scale. Approximately 50% of the shift occurs with a 16-ps time constant and 50% on a significantly longer time scale of 1.8 ns.

The results presented above reveal insight into the structural dynamics, rigidity, and flexibility of oligonucleotides on the picosecond time scale. The differences in the observed magnitudes of the spectral shifts ($\Delta\tilde{\nu}$) suggest that stabilizing chromophore interactions between the two pyrene derivatives are taking place. However, **DNA-16** and **DNA-17** do not show the characteristic pyrene-type excimer fluorescence that is expected to appear at longer wavelengths than the monomer fluorescence. The reason for the absence of this emission is connected with the

Table 9.2 Parameters of the peak-shift correlation functions (PSCF) for **DNA-15**, **DNA-16**, and **DNA-17** obtained from a least-squares fit using a biexponential decay function.

	$\Delta\tilde{\nu}$ (cm^{-1})	τ_1 (ps)	τ_2 (ps)	A_1 (%)	A_2 (%)
DNA-15	440	4.6	220	69	31
DNA-16	1300	7.6	1.9	15	85
DNA-17	1000	1860	16	53	47

a suitable charge acceptor has to be provided for each of the two different charge-transfer processes. Real-time measurements were carried out on ethidium covalently bound to DNA by a flexible molecular tether that enabled intercalation only at specific DNA sites [1]. In these systems ethidium was the hole donor and 7-deazaguanine (Z) the hole acceptor. A remarkable observation was made with this DNA assay: the measured rate constants for charge transfer were independent of the donor-acceptor separation distance in DNA. However, the amplitude of the signal, which reflects the efficiency of the process, decreased with increasing distance. It should be noted that the ethidium/Z redox pair is by now the only DNA-donor-acceptor system that exhibits this unexpected distance dependence of the charge-transfer rate. All other systems have shown an exponential decay of the rate with distance as observed for typical donor-bridge-acceptor systems where the D-A coupling is mediated through off-resonance bridge levels (superexchange).

Methyl viologen (MV) has been used as a noncovalently bound electron acceptor [69, 71]. With respect to the redox potentials, an electron can be transferred from the excited state (E^{*+}) of ethidium to methyl viologen. In most of the corresponding experiments, noncovalently bound ethidium was used to photoinitiate the electron transfer. But the site-specific intercalation of ethidium in DNA is crucial for a detailed study of the binding interactions and the charge donor properties. Thus, we chose a new approach in order to study the photochemical behavior of ethidium in DNA and incorporated **E** as an artificial DNA base at specific sites in duplex DNA [78]. Due to the hydrolytic lability of the corresponding ethidium 2′-deoxyribofuranoside [79], the sugar moiety was replaced by an acyclic linker system that is tethered to the N-5 position of the phenanthridinium heterocycle.

Two arrays of DNA duplexes, **DNA-18–DNA-22** and **DNA-23–DNA-27**, have been prepared using our synthesized **E**-DNA building block. The sequence of this small DNA array was varied in two ways: (1) the base pairs adjacent to the ethidium intercalation site, either G-C or A-T, and (2) the counter base, either T, G, C, or A. Interestingly, all UV-Vis absorption spectra of these duplexes show a maximum in the range of 521–533 nm, which is typical for intercalated ethidium, and there is no significant difference in the absorption maxima (Figure 9.15). Furthermore, the emission maxima can be found in the range of 622–625 nm, which is also typical for intercalated ethidium. These results indicate that the intercalation properties of the ethidium moiety do not depend significantly on the local duplex environment. The intercalation of the artificial DNA base **E** seems not to interfere with the presence of the different counter bases T, G, C, or A. This result is remarkable with respect to the steric demand of the ethidium heterocycle and indicates a bulged position of the counter base.

Temperature-dependent absorption and steady-state fluorescence measurements prove the intercalation of the **E** moiety. The UV-Vis absorption of intact DNA duplexes at room temperature exhibits a maximum of ~530 nm and shifts at higher temperatures to ~505 nm together with a slight increase (Figure 9.16). Furthermore, the intact **E**-modified duplexes show the highest emission quantum yields and a significant fluorescence quenching at higher temperatures. These results clearly indicate that the **E** is intercalated as an artificial DNA and base be-

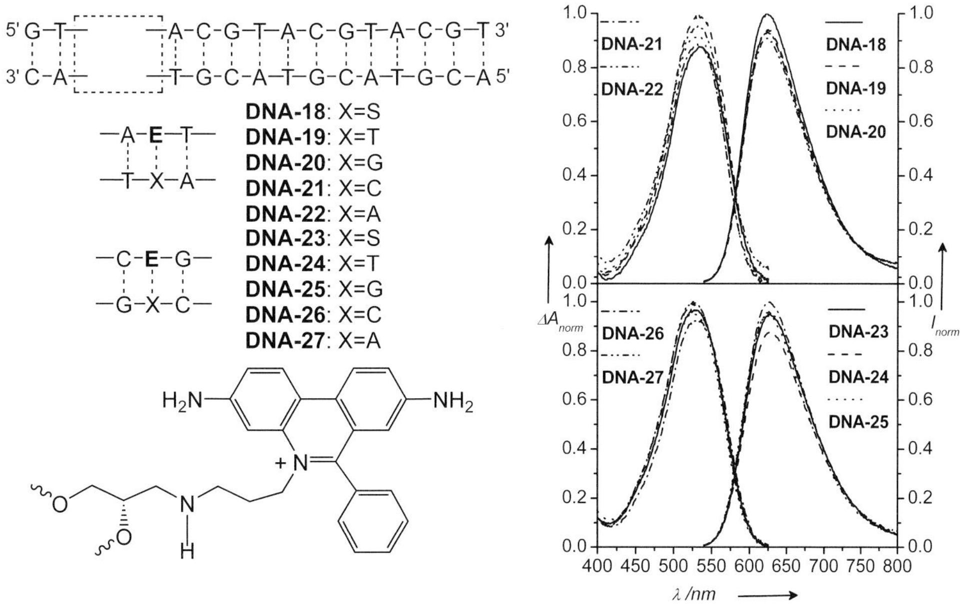

Fig. 9.15 Absorption and emission properties of an array of **E**-modified **DNA-18–DNA-27** (S = abasic site analog).

cause the excited state of the **E** moiety becomes more and more accessible to water at higher temperatures due to dehybridization and base stacking interruption.

The electron-donor properties of **E** as an artificial DNA base were elucidated using methyl viologen (MV) as the noncovalently bound electron acceptor (Figure 9.17) [75]. As expected, the emission of **DNA-18** and **DNA-23** is quenched significantly with an increasing amount of MV as a result of this electron-transfer process. Hence, this **E**-DNA assay has the potential to allow the spectroscopic investigation of reductive electron transfer (not electron hopping) in DNA.

Ethidium in the photoexcited state is also able to photoinitiate DNA-mediated hole transfer when 7-deazaguanine (Z) is provided within the DNA. Two sets of modified DNA duplexes have been prepared: **DNA-30** and **DNA-31** bear the charge acceptor (Z) in a distance of three base pairs away from the charge donor (**E**). A cytosine (C) is placed opposite to **E** in the complementary strand. In contrast, the duplexes **DNA-28** and **DNA-29** lack Z as the charge acceptor and hence provide the control for the assignment of charge-transfer effects. The emission of the photoexcited ethidium is significantly lower when Z is present in the duplex [80]. This observation can be attributed to the charge-transfer process [1, 76, 77]. This interpretation is supported by comparing the amount of emission quenching in duplexes **DNA-28** and **DNA-30**, which is significantly higher than that of **DNA-29** and **DNA-31**. This is due to the fact that a G is located on the 3′ side of Z, which lowers the local oxidation potential, as is known from GG pairs. It is important to point out that the amounts of quenching in both duplex sets indicate a

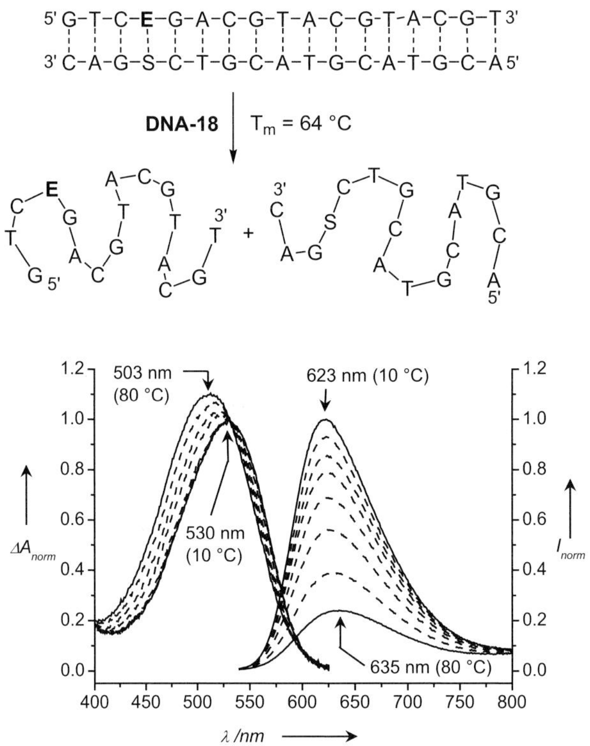

Fig. 9.16 Temperature-dependent absorption and emission of **E**-modified DNA, representatively shown for **DNA-18** (S = abasic site analog).

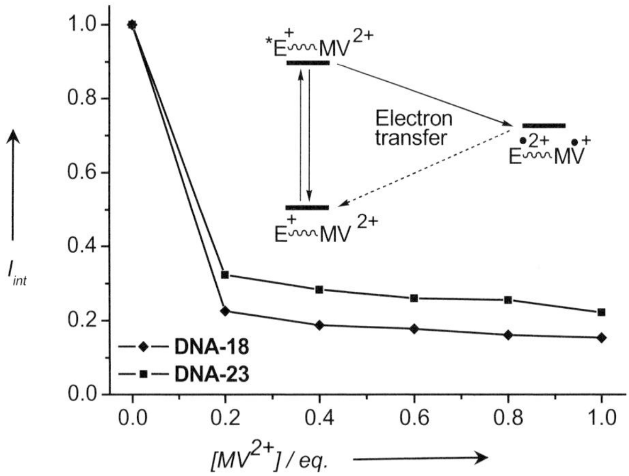

Fig. 9.17 Photoexcited ethidium (E^{*+}) as a charge donor in **DNA-18** and **DNA-23** for electron transfer using methyl viologen (MV) as the non-covalently bound electron acceptor.

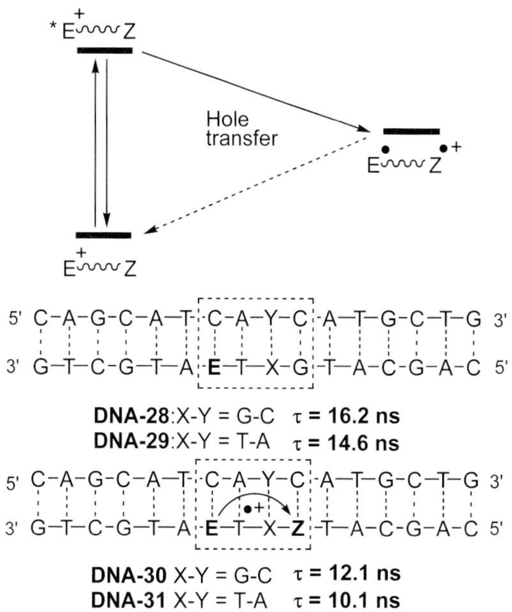

Fig. 9.18 Photoexcited ethidium (E^{*+}) as a charge donor in **DNA-28–DNA-31** for hole transfer using 7-deazaguanin (Z) as the hole acceptor.

quite inefficient charge transfer. In fact, time-resolved measurements gave the fluorescence lifetimes of the four duplexes (Figure 9.18), which yielded charge transfer rates of 28 ns for **DNA-30** and 72 ns for **DNA-31**. These rates are significantly slower than the charge transfer rates of ~5 ps that have been elucidated by using ethidium covalently tethered to the 5′ end of DNA [1]. Interestingly, the earlier measurements showed clearly that the intercalated ethidium needs time (~70 ps) to reorientate before an efficient charge transfer can occur. In our DNA assay, such reorientation could be inhibited as a result of the incorporation of the **E** heterocycle as an artificial base into DNA duplexes. Future experiments will elucidate this unexpected charge-transfer behavior.

9.7 Conclusions and Outlook

Over the last decade there have been many conflicting reports on the electronic conduction properties of DNA. For hole transfer, a broad consensus has been reached as far as the general mechanism is concerned [81]. In contrast, reductive electron transfer has just recently become the subject of intense experimental study. It turns out that both experimental and computational studies are much more difficult to conduct on excess electrons than on electron holes. There are numerous issues that need to be addressed before a comprehensive model can emerge. Some of these issues seem to be equally important for both HT and ET. It has now become clear that the most critical point is the understanding of DNA as a structurally dynamical medium. ET through the base stack cannot be reduced to a static donor-bridge-acceptor problem. Recent experimental and theoretical results have demonstrated that structural dynamics are critical for a comprehensive mechanistic understanding of ET through DNA. In particular, there is strong experimental evidence for an involvement of hydrogen-bond dynamics (not necessarily proton transfer!), which must be considered as a specific mode of solvation dynamics inside the DNA helix. Further experimental studies on carefully chosen functionalized DNA will be needed to obtain a better understanding of DNA dynamics.

Finally, future studies will also have to address the nature of photochemically prepared precursor states in DNA, as their description is relevant for modeling ET rates. Thus, ultrafast electronic energy migration, dissipation, and (de)localization should be convoluted with ET and HT dynamics.

References

1 C. Wan, T. Fiebig, S. O. Kelley, C. Treadway, J. K. Barton, A. H. Zewail, *Proc. Natl. Acad. Sci. USA* **1999**, *96*, 6014.
2 C. Wan, T. Fiebig, O. Schiemann, J. K. Barton, A. H. Zewail, *Proc. Natl. Acad. Sci. USA* **2000**, *97*, 14052.
3 M. A. O'Neill, H. C. Becker, C. Z. Wan, J. K. Barton, A. H. Zewail, *Angew. Chem. Int. Ed.* **2003**, *42*, 5896.
4 M. A. O'Neill, J. K. Barton, *J. Am. Chem. Soc.* **2004**, *126*, 11471.
5 R. Bruinsma, G. Grüner, M. R. D'Orsogna, J. Rudnick, *Phys. Rev. Lett.* **2000**, *85*, 4393.
6 R. N. Barnett, C. L. Cleveland, A. Joy, U. Landman, G. B. Schuster, *Science* **2001**, *294*, 567.
7 A. A. Voityuk, K. Siriwong, N. Rosch, *Phys. Chem. Chem. Phys* **2001**, *3*, 5421.
8 Y. A. Berlin, A. L. Burin, L. D. A. Siebbeles, M. A. Ratner, *J. Phys. Chem. A* **2001**, *105*, 5666.
9 A. Troisi, G. Orlandi, *J. Phys. Chem. B* **2002**, *106*, 2093.
10 A. Troisi, A. Nitzan, M. A. Ratner, *J. Chem. Phys.* **2003**, *119*, 5782.
11 A. A. Voityuk, K. Siriwong, N. Rosch, *Angew. Chem. Int. Ed.* **2004**, *43*, 624.
12 C. E. Crespo-Hernandez, B. Cohen, P. M. Hare, B. Kohler, *Chem. Rev.* **2004**, *104*, 1977.
13 H.-A. Wagenknecht, *Curr. Org. Chem.* **2004**, *8*, 251.
14 M. W. Grinstaff, *Angew. Chem. Int. Ed. Engl.* **1999**, *38*, 3629.
15 F. D. Lewis, Y. Wu, *J. Photochem. Photobiol. C* **2001**, *2*, 1.
16 B. H. Robinson, C. Mailer, G. Drobny, *Annu. Rev. Biophys. Biomol. Struct.* **1997**, *26*, 629.

17 J. Schurr, in *Topics in Fluorescence Spectroscopy, Vol. 3* (Ed.: J. R. Lakowicz), Plenum Press, New York, **1992**, p. 137.
18 E. B. Brauns, M. L. Madaras, R. S. Coleman, C. J. Murphy, M. A. Berg, *Phys. Rev. Lett.* **2002**, *88*, 158101.
19 M. M. Somoza, D. Andreatta, C. J. Murphy, R. S. Coleman, M. A. Berg, *Nucl. Acids Res.* **2004**, *32*, 2494.
20 S. L. Yang, J. S. Cao, *J. Phys. Chem. B* **2001**, *105*, 6536.
21 W. B. Davis, I. Naydenova, R. Haselsberger, A. Ogrodnik, B. Giese, M. E. Michel-Beyerle, *Angew. Chem. Int. Ed. Engl.* **2000**, *39*, 3649.
22 F. D. Lewis, X. Liu, J. Liu, S. E. Miller, R. T. Hayes, M. R. Wasielewski, *Nature* **2000**, *406*, 51.
23 M. Manoharan, K. L. Tivel, M. Zhao, K. Nafisi, T. L. Netzel, *J. Phys. Chem.* **1995**, *99*, 17461.
24 T. L. Netzel, K. Nafisi, J. Headrick, B. E. Eaton, *J. Phys. Chem.* **1995**, *99*, 17948.
25 T. L. Netzel, M. Zhao, K. Nafisi, J. Headrick, M. S. Sigman, B. E. Eaton, *J. Am. Chem. Soc.* **1995**, *117*, 9119.
26 D. O'Connor, V. Y. Shafirovich, N. E. Geacintov, *J. Phys. Chem.* **1994**, *98*, 9831.
27 V. Y. Shafirovich, S. H. Courtney, N. Ya, N. E. Geacintov, *J. Am. Chem. Soc.* **1995**, *117*, 4920.
28 N. Amann, H.-A. Wagenknecht, *Synlett* **2002**, 687.
29 E. Mayer, L. Valis, R. Huber, N. Amann, H.-A. Wagenknecht, *Synthesis* **2003**, 2335.
30 T. Kubota, K. Kano, T. Konse, *Bull. Chem. Soc. Jpn.* **1987**, *60*, 3865.
31 S. Steenken, J. P. Telo, H. M. Novais, L. P. Candeias, *J. Am. Chem. Soc.* **1992**, *114*, 4701.
32 N. Amann, E. Pandurski, T. Fiebig, H.-A. Wagenknecht, *Angew. Chem. Int. Ed.* **2002**, *41*, 2978.
33 M. Raytchev, E. Mayer, N. Amann, H.-A. Wagenknecht, T. Fiebig, *ChemPhysChem* **2004**, *5*, 706.
34 S. Steenken, S. V. Jovanovic, *J. Am. Chem. Soc.* **1997**, *119*, 617.
35 C. A. M. Seidel, A. Schulz, M. H. M. Sauer, *J. Phys. Chem.* **1996**, *100*, 5541.
36 R. Huber, T. Fiebig, H. A. Wagenknecht, *Chem. Commun.* **2003**, 1878.
37 S. Steenken, *Free Rad. Res. Comms.* **1992**, *16*, 349.
38 B. Giese, S. Wessely, *Chem. Commun.* **2001**, 2108.
39 M. Ottolenghi, *Acc. Chem. Res.* **1973**, *6*, 153.
40 E. Pandurski, T. Fiebig, *Chem. Phys. Lett.* **2002**, *357*, 272.
41 D. Mank, M. Raytchev, S. Amthor, C. Lambert, T. Fiebig, *Chem. Phys. Lett.* **2003**, *376*, 201.
42 M. Eigen, *Angew. Chem. Int. Ed.* **1964**, *3*, 1.
43 T. Fiebig, A. Trifonov, H.-A. Wagenknecht, I. Buchvarov, *manuscript submitted* **2004**.
44 M. F. M. Post, J. Langelaar, J. D. W. V. Voorst, *Chem. Phys. Lett.* **1971**, *10*, 468.

45 M. Raytchev, E. Pandurski, I. Buchvarov, C. Modrakowski, T. Fiebig, *J. Phys. Chem. A* **2003**, *107*, 4592.
46 P. Barbara, W. Jarzeba, in *Advances in Photochemistry Vol 15* (Ed.: G.S.H. David H. Volman, Klaus Gollnick (Eds.)), Wiley, **1990**, pp. 1.
47 P. J. Rossky, J. D. Simon, *Nature* **1994**, *370*, 263.
48 T. Renger, R. A. Marcus, *J. Phys. Chem. A* **2003**, *107*, 8404.
49 P. Kaden, E. Mayer-Enthart, A. Trifonov, T. Fiebig, H.-A. Wagenknecht, *Angew. Chem. Int. Ed.* **2005**, *44*, 1636–1639.
50 T. Chen, G. P. Cook, A. T. Koppisch, M. M. Greenberg, *J. Am. Chem. Soc.* **2000**, *122*, 3861.
51 E. Rivera, R. H. Schuler, *J. Phys. Chem.* **1983**, *87*, 3966.
52 T. Ito, S. E. Rokita, *J. Am. Chem. Soc.* **2003**, *125*, 11480.
53 T. Ito, S. E. Rokita, *Angew. Chem. Int. Ed. Engl.* **2004**, *43*, 1839.
54 D. G. McCafferty, D. A. Friesen, E. Danielson, C. G. Wall, M. J. Saderholm, B. W. Erickson, T. J. Meyer, *Proc. Natl. Acad. Sci. USA* **1996**, *93*, 8200.
55 D. J. Freed, L. R. Faulkner, *J. Am. Chem. Soc.* **1972**, *94*, 4790.
56 C. Wagner, H.-A. Wagenknecht, *Chem. Eur. J.* **2005**, *11*, 1871–1876.
57 B. Giese, *Annu. Rev. Biochem.* **2002**, *71*, 51.
58 Z. Cai, X. Li, M. D. Sevilla, *J. Phys. Chem. B* **2002**, *106*, 2755.
59 D. J. Hurley, Y. Tor, *J. Am. Chem. Soc.* **2002**, *124*, 3749.
60 G. T. Hwang, Y. J. Seo, B. H. Kim, *J. Am. Chem. Soc.* **2004**, *126*, 6528.
61 L. H. Thoresen, G.-S. Jiao, W. C. Haaland, M. L. Metzker, K. Burgess, *Chem. Eur. J.* **2003**, *9*, 4603.
62 A. D. Malakhov, E. V. Malakhova, S. V. Kuznitsova, I. V. Grechishnikova, I. A. Prokhorenko, M. V. Skorobogatyi, V. A. Korshun, Y. A. Berlin, *Russ. J. Bioorg. Chem.* **2000**, *26*, 34.
63 V. A. Korshun, I. A. Prokhorenko, S. V. Gontarev, M. V. Skorobogatyi, K. V. Balakin, E. V. Manasova, A. D. Malakhov, Y. A. Berlin, *Nucleosides Nucleotides* **1997**, *16*, 1461.
64 S. M. Freier, K.-H. Altmann, *Nucl. Acids Res.* **1997**, *25*, 4429.
65 A. D. Mesmaeker, R. Häner, P. Martin, H. E. Moser, *Acc. Chem. Res.* **1995**, *28*, 366.
66 M. Rist, N. Amann, H.-A. Wagenknecht, *Eur. J. Org. Chem.* **2003**, 2498.
67 E. Mayer, L. Valis, C. Wagner, M. Rist, N. Amann, H.-A. Wagenknecht, *ChemBioChem* **2004**, *5*, 865.
68 P. Fromherz, B. Rieger, *J. Am. Chem. Soc.* **1986**, *108*, 5361.
69 D. A. Dunn, V. H. Lin, I. E. Kochevar, *Biochemistry* **1992**, *31*, 11620.
70 S. J. Atherton, P. C. Beaumont, *J. Phys. Chem.* **1987**, *91*, 3993.
71 A. M. Brun, A. Harriman, *J. Am. Chem. Soc.* **1992**, *114*, 3656.
72 S. J. Atherton, *J. Phys. Chem.* **1995**, *99*, 12025.
73 A. I. Kononov, E. B. Moroshkina, N. V. Tkachenko, H. Lemmetyninen, *J. Phys. Chem. B* **2001**, *105*, 535.
74 P. T. Henderson, E. Boone, G. B. Schuster, *Helv. Chim. Acta* **2002**, *85*, 135.

75 N. Amann, R. Huber, H.-A. Wagenknecht, *Angew. Chem. Int. Ed. Engl.* **2004**, *43*, 1845.
76 M. A. O'Neill, J. K. Barton, *Top. Curr. Chem.* **2004**, *236*, 67.
77 S. Delaney, J. K. Barton, *J. Org. Chem.* **2003**, *68*, 6475.
78 R. Huber, N. Amann, H.-A. Wagenknecht, *J. Org. Chem.* **2004**, *69*, 744.
79 N. Amann, H.-A. Wagenknecht, *Tetrahedron Lett.* **2003**, *44*, 1685.
80 L. J. P. Latimer, J. S. Lee, *J. Biol. Chem.* **1991**, *266*, 13849.
81 M. Ratner, *Nature* **1999**, *397*, 480.